Patrick Moore's
Practical Astronomy Series

T0222857

Springer
London
Berlin
Heidelberg
New York
Hong Kong
Milan
Paris
Tokyo

Other titles in this series

Navigating
the Night Sky
How to Identify the Stars and Constellations

Guilherme de Almeida

With 105 Figures

Springer

All illustrations, maps and diagrams shown in this book are by the author (except Figure A.1. on page 165).

Translation from the Portuguese language edition of: Roteiro do Céu. Copyright © 1999 by Plátano Edições Técnicas, Lda. Published by arrangement with Plátano Edições Técnicas, Lda. Lisboa, Portugal (3rd edition)

British Library Cataloguing in Publication Data
Navigating the night sky: how to identify the stars and
    constellations. – (Patrick Moore's practical astronomy series)
    1. Stars – Observers' manuals 2. Constellations – Observers' manuals
    I. Almeida, Guilherme de
    523.8
ISBN 1852337370

Library of Congress Cataloging-in-Publication Data
A catalog record for this book is available from the Library of
Congress

Patrick Moore's Practical Astronomy Series ISSN 1617-7185
ISBN 1-85233-737-0 Springer-Verlag London Berlin Heidelberg
Springer-Verlag is a part of Springer Science+Business Media
springeronline.com

Typeset by Expo Holdings, Malaysia
58/3830-543210   Printed on acid-free paper   SPIN 10922033

"I listen and I forget,
I see and I learn,
I do and I understand".

(Chinese proverb)

Note on the Third Edition

The great reception with which the public greeted previous editions of this book has shown that the night sky continues to spark huge interest among people of all ages. It has also encouraged me to retain the original structure of the book. This edition includes various improvements and updates.

Lisbon, November 1999
Guilherme de Almeida

Introduction

Once the Sun has gone down, on clear nights without moonlight, the sky is inhabited by thousands of luminous dots of various colours and levels of brightness. The spectacle offered by a starry sky is both grand and sublime, especially in less urban areas.

In places where night illumination and air pollution are that much more intense, it is well known that we are unable to see as many stars as in a rural region. Nevertheless, this need not be overstated: on the outskirts of cities, or even in the cities themselves, provided you choose the more favourable areas, it is still possible to see many stars.

Navigating the Night Sky is aimed at anyone who marvels at the night sky and who wishes to learn to recognise constellations and identify the brightest stars by name. It is essentially a practical book, which accompanies the reader on their celestial explorations, guiding their way around the stars. Prior knowledge is not required in order to use the book. Anyone can enjoy it, from the youngest student to the interested adult.

There is a popular notion that *all* astronomical observation requires enormous telescopes and profound theoretical knowledge. This is an inaccurate view and may even make people think twice about pursuing this fascinating activity. Even if the reader is unable to identify one single constellation, or any star other than the Sun, please have no fear: you are not the only one in this situation, and in a very short space of time you will no longer be in it. You will see that with this book and with your eyes, even if you have never tried before, you can find your way around the stars. Gradually, almost without noticing, you will pick up the knowledge you need. You will be able to recognise many constellations; you will be amazed by the number of stars you can already identify; you will be able to detect the planets and interpret changes in the overall appearance of the sky, hour by hour and depending on the time of year. So a fascinating adventure unfolds, because the feeling of discovery is exciting and because the knowledge you will pick up from this book is a starting point for other new discoveries.

Where necessary, I repeat certain things, at various points. These repetitions are *intentional*, as the information may be presented in a different context or from an alternative perspective. They will help to clarify certain explanations, make the text generally easier to understand and help to connect the various parts of the book.

The maps in the book and the various indications contained within enable the reader to find their way around the stars, *with the naked eye*. Indeed, you do not learn to recognise constellations or stars and find your way around the firmament with binoculars or with telescopes. It is true that binoculars enable better exploration of the sky and make it possible to see the colour of each star. However, in order to learn to identify the brightest constellations and stars, you have to start with your eyes, the only instrument to offer a full, comprehensive view. In this way you can connect one constellation with another. It is only later that binoculars become appropriate: people learn to walk long before they learn to drive.

Getting to know the sky with the naked eye is *an essential platform* from which to start any astronomical observation. This knowledge will never lose its value, regardless of which binoculars or telescopes you later go on to use.

I hope I have achieved the aims I set for myself. If one clear night the reader feels some sense of contentment and satisfaction when looking at the starlit sky, this book was worth writing.

Lisbon, November 1995
Guilherme de Almeida

Contents

CHAPTER ONE

The Constellations

The first glance at a starry sky shows that the stars are distributed irregularly, more scarcely in some parts, more abundantly in other parts of the heavenly vault. Some are very bright, others are difficult to make out. If you look carefully you can see that stars are also differently coloured: some are yellow, some white, there are some which are redder, some more orange and even some bluish ones.

One gets the feeling that with such variety and disorder it is impossible to find a means of finding your way in the firmament. Many people become disorientated and get lost in the stars. The appearance of the sky changes hour by hour and also depends on the time of year, which only seems to complicate things even further.

It would appear to be impossible to find again, at some future juncture, a star that one has seen on a particular night. It is by no means easy to accept the notion that one can associate stars with others, or give them names, without any risk of confusion. To sum up: the starry sky seems at first to be a complete mess and a challenge for us to detect any kind of meaning in it. Once you overcome these initial difficulties, recognition of the sky is possible, as we shall see. Absolutely anyone is capable of doing this and will gradually make progress.

Thousands of years ago, ancient peoples discovered an effective way of getting to know the sky, although this may *appear* somewhat naïve.

Using great imagination, these peoples realised that the configuration of certain bright stars in the sky resembled an animal, a legendary figure, a hero or something from their daily lives. For example, certain stars seemed to suggest the shape of a bull, others recalled a hunter, some a dolphin jumping out of the water, and so forth. This should not be surprising, given that such names and shapes reflected the times in which they were established.

In view of the fact that the configuration of the stars in relation to each other has apparently remained practically constant throughout many centuries, it is understandable that the shapes imagined in this way and the knowledge associated with them have been passed down from generation to generation, from people to people, across millennia. Such *arbitrary* grouping of the stars into *constellations* has proved useful, as we shall see.

There is, nevertheless, one question hanging in the air: why did the ancient peoples do all of this? Of course, life in these times was less hectic than nowadays and offered plenty of time for reflection and for lengthy contemplation of the sky. Pollution was practically non-existent and there was not the night illumination which we have today in such abundance. The night sky was magnificent and inviting, offering a challenge to more imaginative individuals. But such justifications, while perfectly valid, do not obscure a far more urgent purpose: they *needed* to do it.

Indeed, agriculture, religious festivals and journeys could only be undertaken with knowledge of the sky. Decisions about when to plant seeds and when to harvest, the making of calendars, the best migration times and routes were only possible for those who directed their eyes to the skies and who had adequate knowledge of the constellations.

The constellations which were identified in ancient times were named after animals (Taurus (the bull), Aquila (the eagle), Lupus (the wolf), Ursa Major (The Great Bear), Pisces (the fishes), etc.); hunters who became famous (Orion); heroes and other mythical figures (Hercules, Cepheus, Perseus); musical instruments (Lyra (lyre)); and day-to-day objects (Sagitta (arrow)).

Of course these ancient peoples "found" these and other resemblances *because* these animals and other figures were part and parcel of their daily lives, or of their collective imagination. Some constellations, such as Taurus, Scorpius and Leo were already known in the Valley of the Euphrates, 4000 years before Christ. These were the first to be described in this fashion.

During the second half of the fifteenth century, navigators began to notice that the further south they travelled, the lower Polaris appeared on the northern horizon. Once they were beyond the equator it could no longer be seen. On the southern horizon, stars were becoming visible that had not previously been observed, having been obscured by the horizon (hence not visible in more northerly parts). The configuration of these stars naturally suggested new shapes to these seamen, who began to give them names. Gradually new constellations appeared, some of which were necessary for navigation in these uncharted waters.

After 1603, maps of the stars began to include new constellations. As they originated in relatively recent times, the objects of the day were somewhat different. Thus, gradually, constellations appeared with seafaring names, such as Pyxis (the compass), Circinus (the drafting compass), Sextans (the sextant) and Octans (the octant); the names of other instruments, such as Microscopium (the microscope) and Telescopium (the telescope); the names of birds, such as Pavo (the peacock), Apus (Bird of Paradise) and Tucana (the toucan), as well as various designations of a more technical rather than mythical nature, peculiar to the seventeenth century.

Naturally when we look at the stars today, these stars hardly suggest shapes such as those mentioned above (see Figure 1.1). It would be much more natural for these stars to make us think of, say, a car, a kettle, a telephone or an aeroplane, objects from our daily existence.

Maps of the stars are normally shown in the positive (stars represented as white dots on a black background), or in negative (black dots on a white background). Each of these map types has its upsides and downsides.

In both cases, the size of the "dot" depends on the brightness of the star concerned. Brighter stars are represented by bigger dots and fainter stars by smaller dots. Looking at these maps gives one an idea of the relative brightness of the various stars: if a star is represented by a bigger dot, this is because it is brighter.

Some constellations are easier to recognise than others (see Figure 1.3). Those containing brighter stars, such as Ursa Major, Orion and Scorpius stand out in the sky. Others have fainter stars and are therefore more difficult to detect, such as Lynx and Crater.

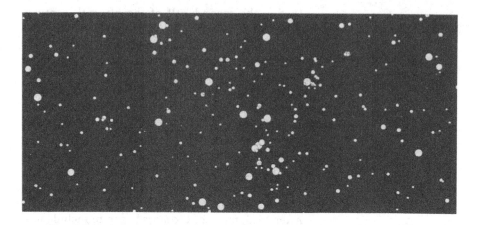

Figure 1.1. Some of the stars visible on a winter's night.

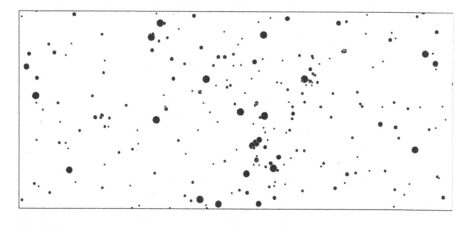

Figure 1.2 The same stars as Figure 1.1 (negative).

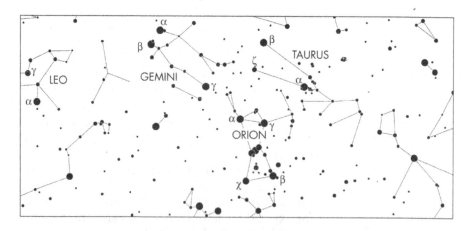

Figure 1.3. The same stars as in Figure 1.1 grouped together in their constellations. The names of some constellations are shown.

Appendix 5 gives an idea of the degree of difficulty of recognising each constellation.

Except in a few cases, there is little or no similarity between the main stars in a constellation and the shapes they are supposed to suggest (Appendix 1 gives some examples). The search for such resemblance may be interesting, but you should try not to be obsessive about it. This is why I am not going to base this study on these shapes, but instead on the gradual recognition of the constellation, beginning with the configuration of the brightest stars.

In order to make it easier to recognise constellations, it is possible to connect their brightest stars by means of lines (see Figure 1.3). The shapes thus revealed can be considered, in some cases, stylised shapes of the ancient representations. They help us to memorise the relative positions of the brightest stars in each constellation, which draw configurations in the sky that are easy to recognise.

1.1 The Viewing Location

On clear nights without moonlight, in non-illuminated areas, the sky is generally very dark, allowing you to observe vast numbers of stars with your own eyes (*naked eye observation*). In such conditions, it is possible to see some 2500 stars at any one time. It is a mesmerising, majestic spectacle. The stars shine intensely above your very head and their presence is unforgettable. In rural areas, away from great population centres, this is how people see the night sky.

However, locations with such *ideal* characteristics are rarely accessible. Most people live in areas that are relatively well lit at night, where it is not possible to see so many stars. Such are the effects of light pollution, an obvious consequence of night-time illumination and a price of progress. There is also air pollution, the result of smoke and exhaust fumes, above your viewing location. The effects of light pollution are exacerbated by air pollution and vice versa: light from street-lamps and other sources is reflected on layers of polluted air and is beamed back to us, giving the night sky a grey, sometimes grey-brown tone. Light pollution is less of a problem when the air is less polluted: air pollution is less harmful if the level of illumination in the area is lower. In most places, the visibility of the stars is mainly affected by light pollution.

If you are in an area with a high level of illumination you will only be able to see the brightest stars. It would be even worse if you were to stand under a street-lamp on a wide avenue. These would constitute *extreme* conditions.

In between extremes, however, there are *satisfactory* conditions, though these may not be ideal. If the area is highly illuminated, you should try to find somewhere less flooded with light. If you are in a city, look to the outskirts. If the place you are in is moderately lit and it is not possible to switch any lights off, you should position yourself in the shadow of a wall or some such obstacle, so that artificial light does not reach your eyes.

Let us not get things out of proportion: in a city, *if* one takes the trouble to avoid less favourable places, it is possible to identify various constellations and stars. On the outskirts of a city, conditions are better still (see Figure 2.3). For beginners, it is better not to have too many stars in view, anyway, as this may be confusing. In such conditions, reasonably bright stars stand out and these represent the most important constellations. Fainter stars, which may get in the way at first, cannot be seen with the naked eye.

It is not difficult to find reasonable conditions: all you have to do is wait for the sky to darken after sunset. As I have said, the brightest stars are the first to become apparent; then, you gradually start to detect other, increasingly faint stars. These are conditions to suit all tastes. In a place with some, but not excessive, illumination, you can find conditions like these and they will stay this way all night. This stage of preference for a "simplified sky" will not last long. You quickly adapt to seeing large numbers of stars without them getting in the way of each other and you will start to explore the magnificent sky that can be seen from the best locations.

If you live in a less favourable place, you can identify the less conspicuous constellations and their most notable stars later on, once you gain a certain amount of experience. Then, make your way to a more favourable location, with this book under your arm. Holidays and weekends may provide the right opportunity for this.

As far as possible, you should seek out somewhere that offers a vast expanse of sky. The ideal scenario, although not essential, would be with an unobstructed horizon, without major elevations or tall buildings.

You should always be comfortably positioned. In order to get a better view of constellations directly above you, it is often a good idea to use a reclining chair or, if you are in a field or on a beach, it is better still to lie on the ground looking up.

The temperature at night is not always as pleasant as one would like it to be. It is therefore important to take enough warm clothing with you.

Last but not least, you should not neglect the question of safety. Not everyone who frequents ill-lit places is there to identify constellations and observe the stars. It is therefore better to go out in a group.

1.2 Adapting the Eyes to Darkness

In order to make the most of observing the sky, your eyes have to be adapted to the darkness. This is the only way you will be able to see the least bright stars accessible to the naked eye. You will be reasonably well adapted 5–10 minutes after you leave an illuminated place, provided you stay in the dark (although adaptation continues to improve for the first 20–30 minutes).

As the process of adaptation goes on, you will see more and more stars, of increasing faintness. The eyes begin to work on *night vision*. This adaptation is necessary, whether you are viewing with the naked eye or using binoculars or a telescope.

There is a simple procedure you can follow to improve still further the capacity of your eyesight. If you are having difficulty detecting a star in a particular part of the sky, look slightly to one side, paying careful attention to the target zone, but without looking directly at it. You can see the star much better and it even looks brighter. But if you move your eyes back to the star it disappears again, because you go back to using a part of your field of vision which is less sensitive to light. This procedure makes use of *averted* or *lateral vision*, which is more sensitive than what you use in the centre of your field of vision.

If you wish to make use of a map, it is better to illuminate this with red light, so that you do not lose your adaptation to darkness. Provided it is not too intense, red light does not dazzle you or make you lose your night vision. If you were to use white light, such as from a torch, you would quickly lose your adaptation to darkness and would have to wait for it to return. One solution is to use a torch covered with a red filter or with red paper, neither too thick nor too transparent.

It is possible to illuminate maps and have both hands free, by using a rectangular torch hanging around your neck on a string.

Of course if you want to learn to identify stars and constellations you have to choose clear nights, without moonlight, as described above. It is recommended to avoid moonlight when you are observing anything in the sky, except when the objects of your observation are the brighter planets or the Moon itself.

In order that the moonlight does not stop you from locating the constellations and identifying the stars, it is important to know how the phases of the Moon are going to affect viewing.

- A *new moon* is not visible, as everybody knows, so this does not present any problem.
- During the *first quarter,* a little patience will go a long way. The Moon goes down around one hour after midnight.
- A *full moon* is visible all night; the intense moonlight "extinguishes" most of the stars.
- During the *last quarter* the Moon only comes up around one hour after midnight, enabling you to observe the sky during the first hours after nightfall.

1.3 What Is the Next Step Towards Identifying Constellations and Stars?

At first glance, constellations appear to be a fixed scene (with the same "design"), which moves around the observer. You can see different parts of the scene, depending on where you are, on the time of night and on the time of year that you are looking at the sky.

The identification of the constellations and the stars is something one acquires little by little, without trying to learn too much on one night (something which should be avoided). Naturally, the amount you learn varies from person to person. Therefore I would suggest that locating *two to five* constellations and identifying their stars (by name, where indicated on maps contained in this book) is enough for one viewing session. More than this and you may begin to get confused. Experience has shown that it is better to learn a little at a time, but assiduously, than the opposite way round.

As mentioned above, it is rare to find a constellation in which the configuration of the brightest stars bears reasonable resemblance to the shapes after which they are named. What may seem such a case of similarity to one person might not be so to someone else. This does not prevent you from recognising constellations; nobody has to "see" the shape of a charioteer in the constellation Auriga, for example, or the Queen of Ethiopia sitting on her throne in the constellation Cassiopeia. In most cases, it is preferable to become accustomed to the configuration of the brightest stars in each constellation, using the Identification Maps (M1 to M8) in this book and later those of the *Celestial Charts* (Maps C1 to C8) as your reference. It will also be useful to *look again* from time to time at the constellations already "discovered", provided they are visible, at the same time as you go on to learn and identify others. You can also start to relate each constellation to neighbouring constellations.

Generally, it is suitable to start learning during the first hours of the night. As the night goes on, further constellations become visible, rising on the eastern horizon, as others dive below the western horizon.

The observation should ideally take place where a large expanse of sky can be seen and where it is easier to connect the various constellations. This kind of sky is also more interesting and offers a magnificent spectacle.

The recommended method is not difficult. You should start by identifying a *small* number of constellations, in different areas of the sky, according to the suggestions in this book. Then following the direction outlined by some bright stars of each constellation you already know, you will begin to make alignments, which in turn enable you to locate other stars and constellations, and these will lead you to further constellations, and so on.

Navigating the Night Sky should not be read in the living-room and later consigned to the book-case. The recommendations in this book are to *be put into practice*. The reader's endeavour and persistence in the quest to become familiar with the night sky will be quickly rewarded.

In this way, the brightest stars and constellations will become familiar. In a short space of time it will be like looking at a street map of your town or one you know well. You know where the main roads lead to and, when you need to, you can find the narrowest streets and all the little nooks and crannies.

If on a particular night you are unable to find a particular constellation or identify a certain bright star, do not despair: *look around* and try to find a constellation you know or a star you can identify. Once you have done this, start your search again from this point, using the Identification Map corresponding to that area of the sky, until you get to the constellation you initially wanted to identify.

Obviously we cannot see the stars in the part of the sky in whose direction the Sun is aligned at certain times of the year (see Figures 3.26–3.28). For this reason, as the year progresses, different stars and constellations are accessible to the observer. It is therefore necessary to have learning sessions during different months of the year. Otherwise you will only get to know stars and constellations visible on one particular occasion.

You can learn even more quickly if, on separate nights, other constellations can be identified some two or three hours before sunrise. In this way, you do not need to wait for months for these regions to become visible at the start of the night.

The maps showing the location of the constellations, which help you to identify the brightest stars, assume that the reader has already learnt how to locate *one* of the constellations contained on each map. Therefore the various maps in this book follow a certain order. On each map, one constellation will be selected as the starting point for new discoveries. From this constellation it will be easy to locate the brightest stars in neighbouring constellations, as we shall see. Once you find a bright star, it will be easy to identify it by referring to one of the maps in this book. From this point on, you will gain some idea of how to locate the constellation to which this star belongs and will be able to recognise it in this area of the sky, based on its three or four brightest stars. For those who live in the Earth's northern hemisphere, one of the best starting points, easily recognisable and viewable all year round, is the constellation known as *Ursa Major*, or the Great Bear.

It is worthwhile knowing the name of the brightest star in each of the constellations that you are going to learn to identify, especially if the star is particularly bright and stands out from those around it. In the case of some constellations with a number of extremely bright stars (of which there are very few), it is worthwhile getting to know the names of two or three of the most conspicuous. You will gradually learn these names, almost without trying. If you observe the sky with the help of the diagrams and maps in this book, you will gradually familiarise yourself with the names of a number of stars that are usually easy to identify.

The basic procedures and knowledge required by the beginner in astronomy can be learnt little by little, following the recommendations given. The more you observe the more you learn. It is worth remembering that one observes for pleasure and for the excitement of discovery, and not out of duty.

Some stars, usually the brightest, have their own names, which tend to be of Arabic or Greek origin. Although all the stars catalogued (over 500,000) have *designations*, we shall only refer to some of those that have their own *names*. As the reader will appreciate, we will direct our attention first towards the brightest stars and to the constellations that are most readily apparent in the sky. Only then will it be possible to begin to consider constellations which only contain stars of fainter brightness levels and which are less easy to identify. In any case, *Navigating the Night Sky* is not intended to be exhaustive.

1.4 Where to Start?

First of all, you need an approximate idea of where north is. You can use Polaris (otherwise known as the North Star or Pole Star) to guide you, if you know where it is. Otherwise, you can position yourself in such a way that the sunset is to your left. This is easy, especially if you ensure you watch at nightfall. Sunrise is to your right and therefore north will be in front of you.

This process will show you approximately where north is, halfway between sunrise and sunset on any given day of the year. You could always use a compass, which will indicate the position of north.

Facing north, once the sky is sufficiently dark, one of the aspects of the sky shown in Figures 1.4–1.7 will be in front of you, depending on the time of year.

It is not difficult to spot the seven brightest stars of the easily-recognised constellation of Ursa Major. Whatever the season in North America or Europe, Ursa Major is one of the constellations that is always visible, assuming the sky is clear. This constellation is therefore a good starting point for learning the sky.

The configuration of the constellations shown in Figures 1.4–1.7 appear to wheel round, either to the left or right, according to the time of year, as evidenced in the maps above. In this way, on the same night, the constellations move round from east to west, approximately around Polaris (in the direction shown by arrows), as the hours go by. Such movement is apparent and is a consequence of the Earth's rotation.

Whatever the time of year in which you begin to explore the sky, these four diagrams will enable you to identify easily three constellations: first Ursa Major, and, in turn, Ursa Minor and Cassiopeia, as we shall see. You can now see that the brightest stars of Cassiopeia draw a letter "M" in the sky, when viewed above Ursa Major (Figure 1.6) and make a "W" when they are below (Figure 1.4).

Figure 1.4.
Configuration of stars in the sky, in **spring**, when the observer looks north at nightfall. Only the brightest stars are indicated. Note the locations of Ursa Major, Ursa Minor and Cassiopeia and the positions of these three constellations in relation to each other. The other constellations in this region of the sky are not shown here.

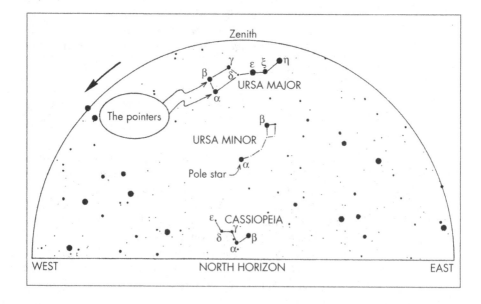

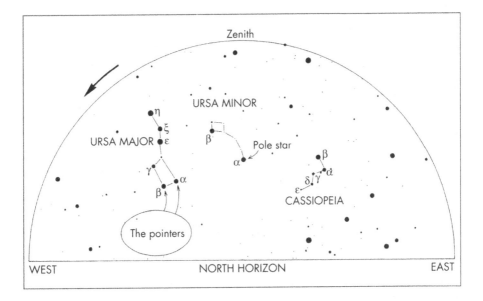

Figure 1.5. Configuration of the stars in the sky, in **summer**, when the observer looks north at nightfall. Only the brightest stars are indicated. Note the locations of Ursa Major, Ursa Minor and Cassiopeia and the positions of these three constellations in relation to each other. The other constellations in this region of the sky are not shown here.

Figure 1.6. Configuration of the stars in the sky, in **autumn**, when the observer looks north at nightfall. Only the brightest stars are indicated. Note the locations of Ursa Major, Ursa Minor and Cassiopeia and the positions of these three constellations in relation to each other. The other constellations in this region of the sky are not shown here.

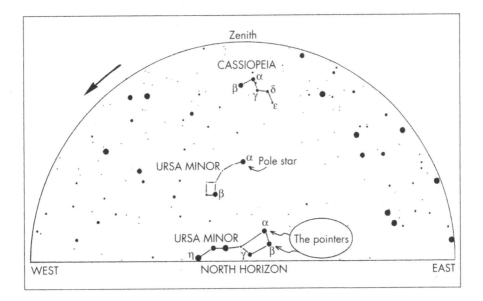

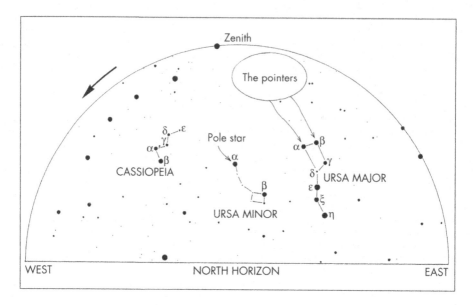

The next diagram, Figure 1.8, is no more complicated than the previous four. You can find your way around, depending on when you begin to observe. This map can also show you Ursa Major, Ursa Minor and Cassiopeia in any season, at any time of the night. The names of certain other constellations have been included, along with the names of some stars. Notice that the two points of the "M" (or "W") of Cassiopeia *are not equal*; the point where Shedar (α Cassiopeiae) is located is at a more acute angle than the other point of the M or W. (Each of these named stars can be found in the Identification Maps in Chapter 4.) This apparently minor detail will later be of use, as it will help you to prevent confusion when you come to use this constellation as a starting point. As regards the constellations close to Ursa Minor, Figure 1.8 shows all of the information in the previous four figures (albeit in more detail). Let us see in more detail how to locate these constellations.

Facing approximately north, hold the book almost vertically, but below eye level, so as not to obscure the sky. Taking Ursa Major as your reference point, turn Figure 1.8 around so that the sky corresponds to this diagram. The stars which are on the bottom side are those adjacent to the north horizon at the moment of viewing. The stars at the top of this map will come up very high in the sky, almost above your head. One map will be enough, so long as it is adjusted according to the month and time it is used. This notion will be

Figure 1.7.
Configuration of the stars in the sky, in **winter**, when the observer looks north at nightfall. Only the brightest stars are indicated. Note the locations of Ursa Major, Ursa Minor and Cassiopeia and the positions of these three constellations in relation to each other. The other constellations in this region of the sky are not shown here.

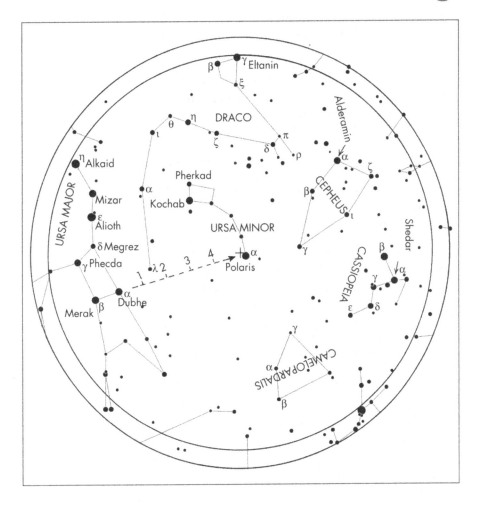

Figure 1.8.

useful when we come to use the *Identification Maps*. These maps should be illuminated with softened red light, so as not to make it difficult for your eyes to adapt to darkness.

As you can see in Figure 1.8, if you follow a line from Merak (β Ursae Majoris) to Dubhe (α Ursae Majoris), the so-called *Pointers* of Ursa Major, and continue five times this apparent distance beyond, you will see a clearly visible star: Polaris (The North Star). This star is the brightest in Ursa Minor and tells you where north is, as follows: if you look at Polaris and then move your eyes *down vertically*, you will find – and this is a reasonable approximation – the north point of the horizon. Note the extension indicated in the diagram is from Merak to Dubhe, in other words in the *opposite* direction to that of the curve of Ursa Major's tail.

Polaris is situated at the tip of Ursa Minor's tail and is not a particularly bright star, but bright enough to make out with little difficulty. As there are no similarly bright stars around it, there is no danger of confusing it with any others, so locating the Polaris is easy. The importance of this star is simply due to its privileged position in the sky, in relation to the observer here on Earth. This current state of *privilege* has not always been the case, nor will it always be. The significance of this will be explored in greater detail in Chapter 3 and Appendix 2.

Once you have identified Polaris, it should be easy to locate two further stars in Ursa Minor: Kochab (β Ursae Minoris) and Pherkad (γ Ursae Minoris). The latter is more difficult to spot with the naked eye if your viewing location is moderately illuminated. You can work out the rest of the configuration in Figure 1.8 from these three stars and see it properly if your place of observation is good. Kochab and Pherkad are sometimes referred to as The Pointers of the Pole. Pherkad is slightly brighter than Megrez (in Ursa Major). Ursa Major and Cassiopeia appear to make the hands of a clock in the sky, with Polaris as the pivot. When Ursa Major is high in the sky, Cassiopeia is very low and close to the horizon, and vice versa. Figures 1.4–1.7 will help you to find this.

Putting these instructions into practice is easier than it may first appear. It is essential to look at the sky itself, as you use this book.

Of course, in any period of around 24 hours at any time of year, the stars of Ursa Major (and those of other constellations) still "go for a walk" around Polaris. Therefore, during the summer months, for example, the winter aspect detailed in Figure 1.7 also occurs, but this is *during the day*, when the Sun makes it impossible to observe the stars.

The configuration drawn by the stars of Ursa Major, connected by lines in Figure 1.8, used to be seen in ancient times as a bear (Appendix 1). Some of these stars are not very bright and thus difficult to see. The seven most apparent stars in this constellation, Alkaid, Mizar, Alioth, Megrez, Phecda, Merak and Dubhe make a shape that is easy to detect in the sky. This is sometimes called *The Big Dipper* or *The Plough*, although in Figures 1.4–1.7 it is referred to as Ursa Major, since these maps only show the brightest stars. An even simpler shape, a quadrilateral formed by Megrez, Phecda, Merak and Dubhe is known as the *bier* or *coffin* (although please bear in mind that these popular designations are not names of constellations).

Chapter 8 shows the aspect of the sky in each of the seasons, regardless of the direction in which you observe.

CHAPTER TWO

The Stars

Stars are very distant suns. The Sun (our star) appears brighter to us because it is relatively close to Earth (around 150 million kilometres). If it were taken much further away, we would see it shining like a star.

Even to the naked eye, we are able to see that the stars display different colours and levels of brightness. As far as brightness is concerned, we can see that some are startling and stand out markedly. Colours range from red to bluish-white, via orange, yellow and even white. This chapter looks at these aspects in greater detail.

The brightest stars acquired their names in ancient times, generally of Greek or Arab origin, and have remained more or less unchanged. Nowadays, there are over 200 stars with their own names and a further 500,000 are catalogued. Of these 200 stars, there are some whose names are commonly known, such as Polaris (α Ursae Minoris), Sirius (α Canis Majoris) and Antares (α Scorpii).

In *Navigating the Night Sky*, around a hundred stars, usually the brightest in the night sky, are mentioned by their own names, and the procedures for identifying these stars are detailed. Appendix 4 gives further useful information about these stars.

Binoculars make stars appear brighter than to the naked eye. They make it easier to recognise colours and to detect stars that are not normally visible to the naked eye. However, through binoculars, or even with the use of a telescope, stars do not appear bigger than to the naked eye.

The observation of an isolated star may not be very interesting, as it will only allow you to appreciate its colour. Groups of stars, on the other hand, particularly large clusters of stars, are extremely interesting and accessible to observation.

Identifying constellations and the brightest stars can be done with the naked eye. Sooner or later, though, once you become familiar with a large number of constellations and can identify various stars, you reach the level where you want to see more. If you do not own a pair of binoculars, this would be the time to consider buying a pair. It is a relatively low-cost investment, which is really useful, even for those who already own a telescope. There are other uses, besides that of astronomical observation. If you already own a pair of binoculars, it might not have occurred to you to point them upwards at a starry sky.

2.1 How Far Away Are the Stars?

Stars are a very long way from us and the distances between them vary greatly from star to star. Looking at the night sky, though, we cannot appreciate the immensity and variety of these distances.

Although astronomical observation dates back millennia and the telescope, at least for the purposes of astronomy, has been around since as long ago as 1609, it was not until 1838 that it was possible to determine the distance between a star and the Earth. This feat was first achieved by Bessel (1784–1846). The star in question was one of the nearest to us and Bessel calculated that it was located around 105,000,000,000,000 km away.

The Sun, which is the nearest star, is 150 million km from the Earth. The next star, in terms of proximity, is 270,000 times further away. In other words, on a scale drawing with the Sun 1 cm from the Earth, the next star would be 2700 m away from our planet.

In the realm of the enormous distances to be found between the stars, the kilometre seems utterly insignificant. To use it for this purpose would be like using millimetres to show distances between the planets of the Solar System, which are in fact tens of millions of kilometres.

Even the dimension of thousands of kilometres is too small for the purposes of astronomy. This is only used to refer to the distance between the Earth and the Moon – 384,000 km on average – and other such distances. Distances within the Solar System are measured in millions of kilometres (the distance between the planet Pluto and the Sun is on average 5900 million km).

For distances within the Solar System astronomers use the *astronomical unit* (AU), which corresponds to the average distance between the Sun and the Earth (around 150,000,000 km). For example, the average distance between Pluto and the Sun is approximately 39.5 AU (i.e. 39.5 × 150,000,000 km) and between Mars and the Sun it is 1.5 AU.

If we wish to show the distance between the stars and the Earth, we have to turn to a vastly greater unit of measurement: *the light year*. One light year is the distance that light travels in space, moving constantly at a speed of 300,000 km per second, during the course of one year. Calculating in terms of the 365.25 days of the year (approximately), the 24 hours of the day and the 3600 seconds in one hour, we can conclude that one light year corresponds to 9,460,000,000,000 km! In other words, in one year, light travels in space a distance 63,240 times greater than the distance from the Earth to the Sun.

After the Sun the nearest star is 4.3 light years away. Even with the naked eye you can see stars located 2000 light years from us. Of course, for a star to be visible to the naked eye from such a huge distance away it has to be extraordinarily bright (dozens of times more luminous than our Sun).

Following this line of reasoning, the light coming from a star which is, say, 500 light years away, took 500 years to arrive here! When you look at this star your eyes are receiving light that it emitted half a millennium ago; you are seeing it as it was 500 years ago: looking into the distance is looking into the past. You will only detect in half a millennium's time any alteration that this star may undergo now.

All the stars that you can see individually belong to our Galaxy, regardless of whether they are viewed with the naked eye, with binoculars or with a small telescope, and even if they are seen from the most favourable locations. Normally known as the Milky Way, this Galaxy is a formidable conglomeration, around 100,000 light years in diameter and made up of some 100,000,000,000 stars and interstellar matter orbiting its nucleus. If your viewing location is sufficiently dark, you will be able to see a milky strip crossing the night sky. This strip, commonly known as the Milky Way, is a view of our Galaxy in profile encircling us.

Bearing in mind the dimensions of our Galaxy, it would be no surprise to learn that the *stars* apparent *to the naked eye* are in our vicinity (in terms of astronomy and given the dimensions of the Galaxy, 2000 light years is not an enormous distance). It would be like saying that in a football stadium you can only see what is two or three metres in front of you.

Chapter 6 contains further information on the Milky Way and on the constellations in whose direction it presents the greatest concentration of stars.

2.2 Are the Stars that Are Most Apparent in the Sky Brighter than the Others?

One interesting characteristic of the stars is their *variety*. Even to the naked eye, we quickly appreciate that these distant suns are differently coloured and of varying degrees of brightness. When it comes to brightness we are easily fooled: an apparently faint star may be highly luminous, but may not seem that way because it is so far away; another shining intensely in the sky, may in fact have only medium or faint brightness, but may stand out due its proximity.

Let us look at an example: in the previous chapter we saw that "Polaris is not an especially bright star". Of course we are referring to the brightness that the star presents to us (apparent brightness). As everyone knows, a source of light, such as a light bulb, appears less bright the further away it is situated. A 150 W bulb 100 m from your eyes seems less bright than a 15 W bulb 10 metres away. The same bulb twice the distance away appears four times less bright, three times the distance nine times less bright, ten times the distance 100 times less bright, and so on.

Looking at two bulbs located at different distances we can say that one of them "is", apparently, brighter than the other. Even so, by simply making a quick calculation, taking into account the respective distances, we can prove that one is actually more intense. This must be borne in mind when referring to the brightness of stars. When you say that a particular star is brighter than another, or that a certain star has a fainter brightness level, you are referring to its *apparent brightness*. This is the brightness that the star *appears to have*, in relation to others, when you look at the sky.

Polaris, which I said above was "not especially bright", is around 430 light years from Earth (27 million times further than the Sun). It is in fact almost 2300 times *brighter* than our star. Seen at a distance of 9 light years, the Sun would show the brightness with which we see Polaris. At 430 light years away, our star could only be seen with a telescope. In Section 2.8, we shall be looking in greater detail at the brightness of the stars.

2.3 Why Do Stars Twinkle?

The Sun does not twinkle, as everyone knows. But we see stars twinkling ("blinking" irregularly). Notice that twinkling is not a characteristic of the stars but merely a consequence of our atmosphere: in order to reach the Earth, the light of the stars has to pass through the Earth's atmosphere. The air is not still, and its temperature is highly variable. Therefore luminous rays will be diverted in various ways, giving the stars this mysterious appearance.

When a star is slightly above the level of the horizon, it twinkles more than when it is higher in the sky. This is because in the former, light has had to pass through a more oblique angle in the atmosphere, passing through a greater air thickness. When you see a star above your head, the twinkling is noticeably reduced (sometimes you can't see it at all), since the light that you receive has passed through less air thickness. On Earth, too, on certain nights, if you are next to a riverbank looking at the light of lamp-posts on the other side of the river, they also twinkle. As it crosses the width of the river, light passes through zones in which the air moves and the temperature varies. This temperature variation gives the air different optical properties and luminous rays. This is why they are subject to slight diversions, which offer a certain irregularity when the light hits your eyes, even if the stars you are looking at have similar brightness.

The light coming from the planets also has to cross the Earth's atmosphere before it reaches your eyes. Generally, though, it is only the stars that twinkle. The stars may be large, but they are so far away that they are just luminous dots, whereas the planets are seen as small circles, even though a telescope is needed for these circles to become perceptible to our eyes.

Some people draw stars with v-shapes or points. You often see pictures like these:

It may seem that the stars look something like these drawings, but this is not borne out by reality. It is easy to do an illuminating and conclusive experiment. All you need is one piece of fine aluminium foil (such as in sweet wrappers). Make a small hole using the point of a needle, bring the hole close to your eye and look at a shining star. What you see is a luminous dot, not pointed at all. Using this hole, the light entered your eye only through the crystalline centre – the "eye's lens". Without looking through such a hole, light would pass through the centre *and* via peripheral parts of the eye, whose structure has certain radial imperfections on the outer edges. In other words, the explanation lies in our eyes and not in the stars, which are practically spherical. Foil is used because with paper it is difficult to make a perfect hole without a "furry" edge.

2.4 Do the Stars Move?

All the stars in our Galaxy (the *Milky Way*) move, orbiting around the galactic nucleus at fairly high speed; the orbital speed of the Sun, for example, is around 250 km a second. The stars nearest to the galactic nucleus complete one lap in less time than those further away. In Chapter 6, you will read in more detail about our Galaxy and the constellations where its characteristic whitish strip passes through.

Measuring the speed of the stars in relation to the Sun, it has been observed that speeds have slight variations from star to star, but are measured in tens of km per second. However, since the stars are a great distance from us, their changes of position in the sky (some moving further and more quickly than others) only become visible to the naked eye after several thousand years. Therefore we can consider that the "shape" of each constellation has remained fairly constant throughout many generations. It is also true that in terms of longer periods of time the "shape" of the constellations will undeniably change as time goes on.

2.5 What is the Size and Mass of a Star?

The sizes of the stars vary greatly from star to star. The Sun, the only one that can be seen relatively close up, is, on average, 1,400,000 km in diameter, which is still more than 3.5 times the distance from the Earth to the Moon. The diameter of a star ranges from a few thousand km, in the case of the dwarfs, to more than 600 solar diameters, in the case of the supergiants. Note that the "diameter" of the Earth's orbit around the Sun is around 215 solar diameters, which gives you some idea of just how enormous some stars are.

Are the stars heavy? Before I answer this question, it is worth pointing out that physicists and astronomers, who like to be absolutely clear about what they say, refer to size measured in kilograms as *mass*, not weight. It is also true that an object of greater mass is also heavier.

Now that we are using the term "mass" properly, we can say that in this regard the stars vary enormously. The Sun, an average star in terms of mass, is 1,990,000,000,000,000,000,000,000,000,000 kg, which is 330,000 times the mass of our planet!

In order to work, i.e. to emit radiation, a star must have at least 7% of the Sun's mass (0.07 of the solar mass). At the other end of the scale, supergiant stars can be up to 60 solar masses.

2.6 How are Stars Formed and How Long Do they Live?

Stars are enormous balls made up mainly of hydrogen. The Sun is much nearer to us than all the other stars and therefore displays a dazzling shine. Stars are formed by the agglomeration of immense clouds of dust and gas in interstellar space, which are gradually compacted due to gravity. In this way, the matter later to become the star is compacted increasingly intensely, raising the temperature inside so high (10,000,000 °C) that atomic nuclei of hydrogen are able to combine, thus making helium nuclei. This process is known as nuclear fusion and releases energy: the star begins to shine. In these conditions the compression stops and the star enters a relatively stable and long-lasting stage of its life.

While these reactions are taking place, the temperature inside the star goes up and the hydrogen used initially gradually dwindles. At higher temperatures, other nuclear reactions take place involving heavier elements (carbon, nitrogen and oxygen, etc.) and these are the direct result of previous nuclear reactions. These new reactions, in turn, increase still further the temperature of the star's nucleus.

Using various different types of "fuel", which are gradually exhausted and make way for others which, in turn, cause reactions that enable the star to be even hotter and even brighter, the star begins to waste away: stars do not live forever. The greater the amount of matter contained in the star at the outset (greater mass), the quicker it will consume its fuel, so the brighter it will be and the shorter its life will be. The life span of a star ranges from a few million years, in the case of those with the greatest mass, to hundreds of thousands of millions of years, if the mass is sufficiently low. Stars such as our Sun live around ten thousand million years; our star has already lived around half of this length of time.

However, the amount of matter from which a star is formed does not determine solely the greater or lesser length of its existence, it also affects the way in which it will end its life. If the mass is relatively small, it will cool down gradually, without ever shining intensely. It may swell like a giant red balloon, and later, in the final throes, release part of the matter into space and then contract, making a very compact star (white dwarf), which will slowly cool down. There may be a tremendous explosion engendering a small and highly concentrated star (a neutron star) or in extreme cases, giving rise to an extreme concentration of matter, subject to its own gravity to such a degree that light is unable to emerge (a black hole).

Throughout its life a star will produce, through nuclear fusion, nuclei of heavy elements from hydrogen and helium. Almost all the known elements originate in this way, in high mass stars. The stars are "machines" producing atomic nuclei. Without them there would only be lightweight elements in the Universe. Carbon, iron, calcium and the other elements that are an integral part of our bodies and all objects around us, were forged in the stars of previous generations, which, on exploding, disseminated new elements and enriched the Universe.

2.7 The Colours of the Stars

At first glance, all stars appear white. But further inspection reveals that some are reddish orange, such as *Betelgeuse* (α Orionis), *Antares* (α Scorpii), and Aldebaran (α Tauri). Others are a shade of bluish-white, such as *Rigel* (β Orionis) and *Spica* (α Virginis). There are also yellow stars, such as *Capella* (α Aurigae) and our Sun, while others are white, such as *Sirius* (α Canis Majoris) and *Altair* (α Aquilae).

Throughout its life each star goes through different stages of evolution, during which the "surface" temperature changes. Red stars are already in the final stage of their existence.

There is a correlation between the colours of stars and their surface temperatures. The higher the surface temperature of a star, the more its colour will be bluish. Stars with lower temperatures are redder in appearance. The examples in Table 2.1 give some idea of the relationship between the colours of stars and the corresponding approximate surface temperatures.

The nucleus of a star is extremely hot, with temperatures reaching up to millions of degrees Celsius. In the

Table 2.1. Colours and surface temperatures of stars

Colour	Surface temperature	Example star	Constellation
Red	3000 °C	Betelgeuse	Orion
Orange	4000 °C	Arcturus	Boötes
Yellow	6000 °C	**Sun**	
Yellow-white	8000 °C	Procyon	Canis Minor
White	11,000 °C	Sirius	Canis Major
Blue-white	25,000 °C or above	Spica	Virgo

Sun's nucleus, for example, the temperature is some 15,000,000 °C.

2.8 The Brightness of the Stars

It is well known that when we look at the sky we see stars of varying colours and brightness. They show different levels of brightness for two reasons. Firstly, because they simply do not have the same luminosity: while some stars have less than 0.0004% of our star's luminosity, others shine brighter than 150,000 suns. Secondly, because the stars are at different distances from us: even if we limit ourselves to those which can be seen by the naked eye, some are hundreds of times further from us than others.

You can find further information on the bright stars that appear in this book's maps in Appendix 4.

In order to make the best use of the maps and to have a better idea of the brightness displayed by the corresponding stars it is useful, *even essential*, to have a notion of the classification of the brightness of the stars.

Previously, such classification was based on *greatness*: the brightest stars were said to be *first greatness*, other less bright stars were considered *second greatness* and so forth. The stars which were the least bright to the naked eye (in an excellent viewing location) were *sixth greatness*. As this classification referred (and still refers) to *apparent brightness*, and had nothing to do with the actual size of the stars, it was abandoned (as recently as the nineteenth century), so as not to cause confusion. So people stopped using *greatness* and began to say *magnitude*. This was more than a mere change of words – the numerical value of each star's magnitude began to be measured rigorously and defined precisely, which led to some changes in the "league table".

Currently, the apparent brightness of stars is classified in a numerical scale, known as *the magnitude scale*. The *lower* the brightness displayed by a star, the *greater* the numerical value attached to its brightness (the greater its magnitude). Table 2.2 gives an idea of the brightness of stars visible to the naked eye, according to corresponding magnitudes.

It is assumed for the purposes of assessing the conditions referred to in Table 2.2 (especially for the less bright stars) that the observer's eyes are already adapted to the darkness. From magnitude 6 to magnitude 1 the brightness is increasingly high. In this way, the greater the brightness displayed by a star, the lower its magnitude value (a second-magnitude star, for example, looks brighter than a third-magnitude star).

Figures 2.1 and 2.2 show the magnitude of certain stars which are easy to identify. Looking at these figures and at the corresponding stars in the sky, the reader has a number of examples which will serve as a reference for estimating the magnitudes of other stars. The maps in this book also offer some idea of the brightness of the stars: all you have to do is look at the size of the dots – the bigger dots represent brighter stars.

If you are in a place with street lamps or other intrusive sources of light which cannot be switched off, visibility will improve appreciably if you sit behind a wall or other obstacle which puts you in the "shade" of this light. As mentioned above, night illumination and air pollution make it difficult to observe the less bright stars. In a moderately illuminated urban location you can only see with the naked eye stars brighter than magnitude 3.5. The Identification Maps show all stars up to magnitude 4.8, corresponding to what can be seen in a reasonable location. Figure 2.3 gives examples of visibility in various different viewing conditions with the naked eye.

Regardless of the "purity" of the air, viewing conditions can vary, according to the transparency of the atmosphere and humidity levels at the time and place of observation. If the Moon is visible, it will be impossible to see the dark sky or the less bright stars which would be apparent in more favourable conditions. The

Table 2.2. Magnitudes of stars and perception of brightness

Magnitude	Observations
Sixth magnitude (or magnitude 6)	Very faint stars. Difficult to see even for those with excellent eyesight, in a *perfect* location, such as in the countryside, very dark and with little pollution, on a clear night. At the limits of visibility to the naked eye.
Fifth magnitude (or magnitude 5)	Slightly brighter than sixth-magnitude stars. These stars are at the limits of visibility to the naked eye if the observer is in a *good* location, a reasonable distance away from the city.
Fourth magnitude (or magnitude 4)	Stars visible to the naked eye if the viewing conditions are simply adequate. In a city, even in a moderately illuminated location these stars are not usually visible. On the outskirts of cities, these stars will be visible if the location chosen is sheltered from light pollution.
Third magnitude (or magnitude 3)	Stars easily seen with the naked eye. Viewable even in cities. Only difficult to see if the conditions are extreme: intense illumination and highly polluted air. The star Megrez (in the constellation Ursa Major) is a perfect example of the brightness corresponding to third magnitude.
Second magnitude (or magnitude 2)	Stars easily seen with the naked eye. Polaris is a perfect example of the brightness corresponding to second magnitude.
First magnitude (or magnitude 1)	Stars which stand out in the sky with their brightness. The first stars to appear when night falls. The star Spica (in the constellation Virgo) is a perfect example of the brightness corresponding to first magnitude.

observation of less bright celestial bodies will be affected. The more luminous planets can still, however, be clearly seen. On nights when there is a full moon, only stars with a magnitude of less than 3 can be seen, regardless of the viewing location.

Not all stars apparent in the sky are first magnitude ($m = 1$). Therefore, those of a brighter magnitude than 1 are said to be magnitude zero ($m = 0$), as exemplified

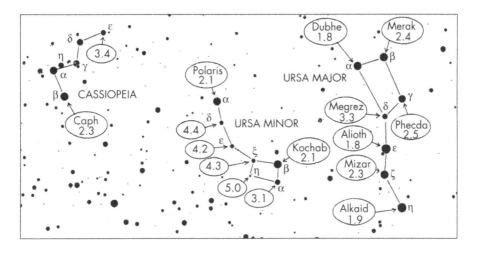

by the star *Vega* (α Lyr) in the constellation Lyra. Those with an even greater brightness are still referred to as magnitude –1 ($m = -1$). There is only one star brighter than magnitude –1: *Sirius* (α Canis Majoris), in the constellation Canis Major (see Figure 2.2).

A star with a medium brightness between magnitudes 2 and 3 is brighter than one in magnitude 3 and less bright than one in magnitude 2. If a star's brightness is at the halfway point between the two, it is said to have a magnitude of 2.5. This gives you an idea of the meaning of, say, $m = 1.6$ or $m = 3.7$.

You can use some stars to evaluate conditions in your viewing location (see Figure 2.1). An observer in a city, who has chosen a location protected from light pollution, can only usually see two stars in Ursa Minor with the naked eye: *Polaris* and *Kochab*; in slightly

Figure 2.1. Examples of magnitudes of some stars in the vicinity of Polaris.

Figure 2.2. Examples of magnitudes of bright stars and of less bright stars in the constellations of Canis Major, Orion and Canis Minor. These constellations are identified in Map M3.

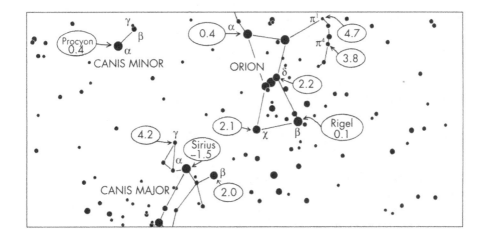

better conditions one would see the third brightest star in the constellation (Pherkad).If the viewing location is very good, you will still only see the seven brightest stars of Ursa Minor, as shown in the diagram. This is therefore a useful and accurate means of testing the conditions.

As far as Ursa Major is concerned, if viewing conditions are inadequate, the star Megrez (magnitude 3.3) will not be visible to the naked eye. If conditions are reasonable, then you will clearly be able to see the seven brightest stars in the constellation.

Only three stars are brighter than magnitude zero: *Sirius* (α Canis Majoris); *Canopus* (α Carinæ, the keel), which can only be seen from latitudes south of 33° (southern USA, not at all from Europe), and *Alpha Centauri*, which cannot be seen from Europe or the USA. It is usual to consider magnitude classes, or rounded magnitudes, as seen in Table 2.3; in this way, for example, stars with a magnitude between 1.50 and 2.49 are referred to as second magnitude.

The number of observable stars grows considerably if more faint stars are taken into account (greater m), as can be seen in Figure 2.3. In the entire celestial sphere, in the best viewing conditions, only around 6000 stars are visible to the naked eye.

The light that arrives here from celestial bodies just above the horizon has to pass through the atmosphere very obliquely, thus crossing a greater thickness of air. This light is therefore weakened to a greater extent than the light of bodies that are actually above our heads. Stars that are only slightly above the horizon look less bright. This *weakening* process is gradual and more evident the closer the star is to the horizon at the time of observation. At less than 10° it is significant. In the case of the Moon, this phenomenon is obvious, as readers will no doubt have noticed. This is why, at any given moment, an observer with good eyesight, who is observing at an *excellent* location, will only be able to see around 2600 stars (less than half the 6000 mentioned previously). It should also be noted that Table 2.3 includes stars up to magnitude 6.49, whereas our eyes only "reach" sixth magnitude, in the best conditions, without moonlight.

I mentioned earlier that while a star may appear brighter this does not necessarily mean that it is in fact brighter. The apparent magnitude of a star only indicates the brightness that it displays, seen from Earth, and does not refer to the fact or otherwise of its luminosity.

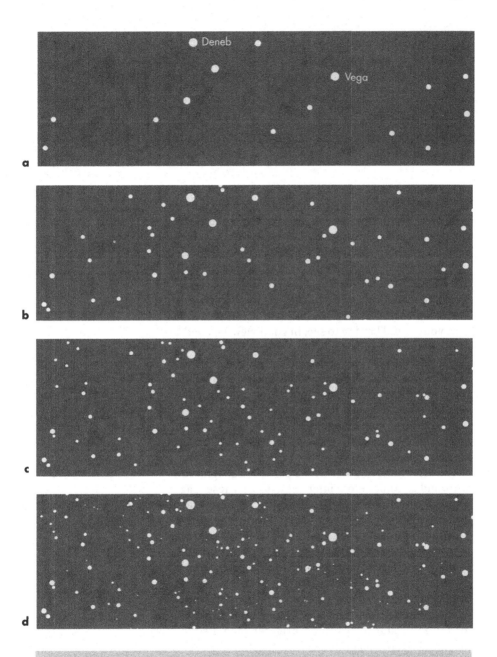

Figure 2.3. The same part of the sky, seen in different viewing conditions: in an illuminated area of a city (**a**), in an area protected from illumination, also in a city (**b**), on the outskirts of a city (**c**) and in a dark area, away from light pollution (**d**). The smallest binoculars will show much more. The four images are all centred on the constellations of Cygnus and Lyra, whose brightest stars are Deneb and Vega, respectively.

Table 2.3. Rounded magnitudes

Brightness between magnitudes	Rounded magnitude	Number of stars observable in the whole sky
Less than 1.49	first magnitude	20
Between 1.50 and 2.49	second magnitude	53
Between 2.50 and 3.49	third magnitude	157
Between 3.50 and 4.49	fourth magnitude	506
Between 4.50 and 5.49	fifth magnitude	1740
Between 5.50 and 6.49	sixth magnitude	5170
Etc.	Etc.	

The use of binoculars massively increases the possibilities for observing the stars and other bodies. The stars become brighter, stars that were bright before become brighter still, those that were difficult to see with the naked eye are now obvious, thousands of stars that cannot be seen with the naked eye are now accessible, all of which increases the interest in observing. If the "front lenses" of the binoculars are 50 mm in diameter, you should be able to see, in good viewing conditions, magnitude 10 stars. More than 100,000 stars become accessible. Telescopes offer even greater possibilities. This capacity increases in proportion to the size of the lens on the instrument used. A telescope with a 100 mm diameter lens, or mirror, enables observation of stars up to a magnitude of 11.8. Nowadays, in major observatories, it is possible to make out stars up to magnitude 25.

The brightness of a star with a particular magnitude is around 2.5 times more intense than one whose magnitude is one unit above. For example, a second-magnitude star appears 2.5 times brighter than a third-magnitude star. This in turn is 2.5 times brighter than a magnitude 4 star and so forth. (This amount is in fact 2.512, but rounded down to 2.5 here.) Taking this scale further, a fifth-magnitude star is 6.3 times brighter (2.512×2.512) than a magnitude 7 star, which is now past the point of visibility to the naked eye. A magnitude 1 star is 100 times brighter than a sixth-magnitude star (5 magnitudes difference).

This scale is also applicable to the planets and their satellites, including our Moon. Although they do not have their own light and only reflect what they receive from the Sun, they still have a certain degree of brightness, depending on the viewing location. Naturally, this scale also applies to the Sun. Table 2.4 shows apparent magnitudes of the Sun, Moon and some other planets.

Table 2.4. Magnitudes of the Sun, Moon and some other planets

Body	Magnitude
Sun (zenith)	−26.8
Full moon	−12.6
Quarter moon	−9
Venus (maximum brightness)	−4.3
Mars (maximum brightness)	−2.8
Mercury (maximum brightness)	−1.9
Jupiter (maximum brightness)	−2
Saturn (maximum brightness)	−0.4
Uranus (maximum brightness)	5.7

The Sun is the only star we can see close up. However, one should not look at our star, even with the naked eye, without adequate protection, under any circumstances. One should never, for any reason, observe the Sun through binoculars or telescopes.

If it were very far away, say, 9 light years, the Sun would shine with the same brightness as Polaris (second magnitude). If we saw it 57 light years away it would be at the limit of visibility to the naked eye (magnitude 6).

2.9 Do Stars Always Have the Same Brightness?

As a star evolves, its brightness changes. It is usually, however, an extremely slow process, which would be undetectable over centuries of observation, let alone during the lifetime of a human being. The brightness of the Sun, for example, is little different from what it was 10,000 years ago. It is like watching a child – we do not notice its growth over the course of a few seconds. This is therefore not a problem we need to deal with as far as naked eye observation is concerned, except in the case of a star that ends its life with a violent explosion (supernova) and is thousands of light years from Earth. This happens with large mass stars, whose brightness in such conditions increases a million times. It is extremely rare – in each galaxy it happens on average less than once every two centuries.

Nevertheless, some stars do show much quicker variations in brightness, which generally have nothing to do with the evolution of the star from one stage to the next. These are known as *variable stars*. It is not always a case of genuine alterations in brightness. In some cases they are double stars in which the components have a constant brightness. The component star with the smaller mass orbits the other (as is the case of the Earth, which orbits the Sun). If such an orbit is seen from the Earth in profile, the satellite star passes from time to time in front of the main star, obscuring it for a while and giving the impression that its brightness varies regularly. In fact this is an eclipse and the stars in question are referred to as *eclipsing variables*. The star Algol, in the constellation Perseus (Map M4), is a classic example of an eclipsing variable. The minimum brightness stage lasts about ten hours.

In other cases the star itself varies in brightness, due to temporary imbalances between energy radiation (which tends to make the star swell) and gravity (which tends to compress it). In these conditions, the star pulses and varies in size and brightness. Such stars are naturally variable and can be of various types. Depending on the type, variations in brightness are regular, semi-regular or irregular. Periods of variation can also diverge substantially, ranging from a few days or weeks (regular) to months or even years (irregular). The star Mira, in the constellation Cetus (Map M4), is an example of a long-period variable (331 days). Its diameter oscillates periodically between 300 and 400 times that of the Sun. The star Betelgeuse, a red giant in the constellation Orion (Map M3), is a semi-regular variable. Its brightness varies between magnitudes 0.7 and 0.4 during a period of approximately five and a half years.

There are also some double stars whose components have very different masses, in which case the star with the greater mass, having evolved more quickly, is said to be in a stage of evolution called a "white dwarf". It attracts outer layers from the accompanying star, giving rise to violent explosions, which raise its brightness for days, weeks or even months. This is a rare phenomenon and the stars which behave in this way are known as "*novas*" even though they are obviously existing stars that have become temporarily more noticeable.

It is clear that in this chapter it would have been impossible not to mention, albeit briefly, stars of this kind. However, it also true that the observation of variables is outside the remit of this book and the number of variables with an appreciable level of variation visible to the naked eye is extremely low. The existence of variable stars does not affect the recognition of stars and constellations. Observing variables requires good knowledge of the sky and much persistence.

2.10 "Stars" that are not Stars

On certain dates, at the same time of year every year, one can see luminous traces flashing across the night sky. In most cases, these last less than a second and are usually referred to as *falling stars*, although astronomers call them *meteors*. They can seem to be falling from the sky, but *have nothing to do with stars*. They are in fact small particles, usually less than one gram in weight, which penetrate the Earth's atmosphere. They were left behind by comets, on the way around their orbits. When the Earth crosses a comet's orbit, it attracts these particles and many of them are hurled down to our planet at great speed (around 60 km/s), catching fire through friction when they cross the Earth's atmosphere. They turn incandescent (between 120 and 50 km in altitude), emit light and usually vaporise completely, not reaching the ground. When there is a large number of them, we call this a *falling star shower* or *meteor shower*. Most of the meteors produce a trace of (apparent) brightness in the sky less than that of a first-magnitude star. However, there are times when much brighter meteors can be seen, since particles with greater mass cause brighter traces. Every year the dates when this occurs most, each linked to a particular comet, are the nights of 11–12 August (50 meteors per hour, on average), 21–22 October (25 meteors per hour), 13–14 December (50 meteors per hour) and 3–4 January (40 meteors per hour). They are also visible some days before and after these dates, but with fewer meteors per hour.

Comets only become reasonably bright when they pass in their orbits closest to the Sun (perihelion). However, comets that are bright enough to be seen with the naked eye are relatively rare occurrences. For this reason, although meteors are related to them, comets do not play a major role in this book.

CHAPTER THREE

The Celestial Sphere

When you look at a starry sky, you have the *feeling* that where you are standing is at the centre of a giant hollow ball (Figure 3.1). Half of this spherical surface can be seen above the horizon, like a heavenly roof. All the stars appear to be "glued" to the internal surface of this massive sphere, giving the impression that they are all the same distance away. This feeling is enhanced if the viewing location is unimpeded and encompasses a large expanse of sky.

You know that the stars are spread out at various distances away, but as you look at the heavenly roof it

Figure 3.1. An observer has the feeling that the stars are fixed to a heavenly roof and it appears that the point of observation is immediately below the centre of this roof.

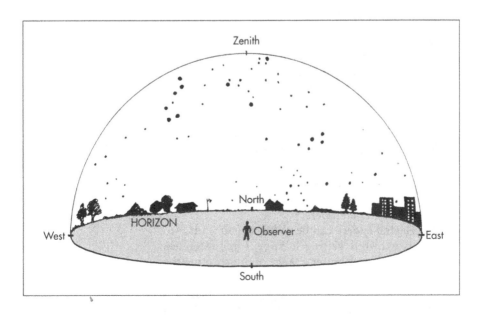

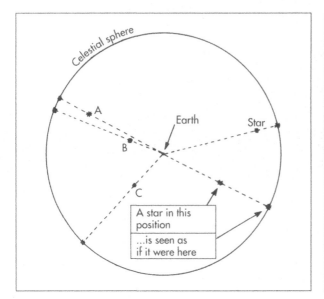

Figure 3.2. Regardless of their various distances from the Earth the stars seem to be projected onto the celestial sphere and to be the same distance away from us. Stars A and B seem to be side by side, but they are not the same distance from Earth, and are in fact a long way from each other. The Sun is not represented here, as it is relatively close to the Earth. This diagram is not drawn to scale, as that would be impossible.

is hard to appreciate this (see Figure 3.2). In the same way, although it is common knowledge that the Earth is more or less round and that it rotates on its own axis, it still looks flat and stationary.

As the hours go by the stars seem to move from east to west, which, as we know, is merely a consequence of the west to east rotation of the Earth. This phenomenon is like being on a merry-go-round moving anti-clockwise; it appears when you look outside that the rest of the scene is constantly moving to your right. If the merry-go-round is uncovered, the clouds appear to be spinning on the merry-go-round's axis, in the opposite direction to the one in which it is moving.

The idea that the stars move around the Earth is based on appearances: it is false, as we have seen, and has nothing to do with reality. Nevertheless, in order to identify the stars and the constellations and to understand the hourly and nightly alterations in the sky's appearance, it is useful to proceed *as if* this immense sphere existed in reality. In astronomy this sphere is known as the *celestial sphere*.

The celestial sphere can be as big as you want it to be. It is worthwhile getting to know some reference points, to give you a better chance of interpreting the sky's appearance.

Let us suppose you are standing up. Try to imagine a vertical straight line from your feet up to your head and

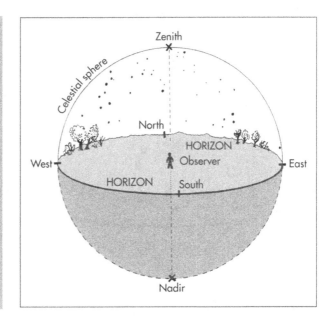

Figure 3.3. Horizon, zenith and nadir, along with the points of the compass north, south, east and west.

this line carries on straight up past your head, and onwards to a point in the celestial sphere. This point is known as the zenith (see Figures 3.3 and 3.4). Slightly more difficult to imagine, but the continuation of this

Figure 3.4. Some important reference directions for an observer situated in a particular location. Due to the insignificance of the dimensions of the planet Earth in relation to the celestial sphere, the observer looks towards the celestial north pole, in a parallel direction to the axis of the Earth's rotation. Another observer in the southern hemisphere will look towards the celestial south pole in a parallel direction to the same axis.

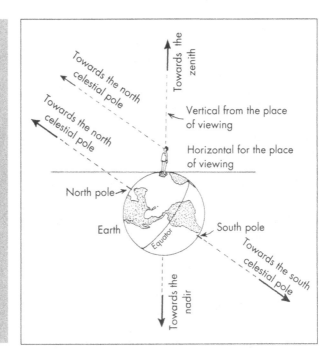

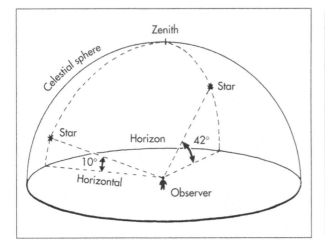

Figure 3.5. The altitude of a celestial object. The star on the left is at an altitude of 10°; the altitude of the star on the right is 42°. As mentioned in the text, the altitude of a star depends on the time and place of observation.

line in the opposite direction down below your feet and onwards through the Earth reaches the opposite point to the first. This point is called the *nadir*. It may seem strange but there is sky below us on the underside of the globe. Around us is the horizon, whose outline is often altered by mountains, trees and buildings. This is where the basic compass points, north, south, east and west are situated.

Whenever you look at a clear sky you can see stars that are almost at the zenith and others close to the horizon. The *altitude* of a celestial body is the angle between the horizontal and the direction in which the body is observed (see Figure 3.5). So, the altitude of a star at the horizon is said to be at 0°; at the zenith this figure is 90°; and at the half-way point between the two the altitude is 45°. This applies to any celestial object. We can also say that the *altitude of the zenith* is 90°.

The position of a star in relation to the horizon, and to these base points, changes constantly. The celestial sphere rotates at around 15° per hour. This apparent rotation, as has already been mentioned, is due to the Earth's rotation.

The Earth's equator is, as we know, perpendicular to our planet's axis of rotation. If you imagine that the equator were to continue outwards through space, it would eventually intersect the celestial sphere. This line is known as the celestial equator (Figure 3.6) and divides the celestial sphere into two hemispheres –

Figure 3.6. The celestial equator and the celestial poles.

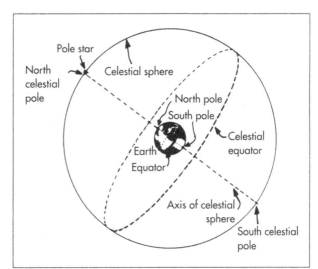

north and south. The celestial equator can also be described as a projection of the Earth's equator in the celestial sphere (equivalent to the definition above).

If the reader imagines the axis of the Earth's rotation extending from both sides of the globe, the line will be in the direction of the celestial poles (see Figures 3.4 and 3.6). The northward extension points toward the celestial north pole, very close to the direction in which we observe Polaris, which is currently at around 0.8° from the celestial north pole. The southward extension points toward the celestial south pole in a region of the celestial sphere where there is no especially bright star that could feasibly play the role of southern pole star.

The celestial south pole is not visible from Europe or the USA, as it is situated below the horizon (see Figures 3.15 and 3.16). It is also true that from a location situated at 40° N it is impossible to see stars at less than 40° from the celestial south pole, as they are below our horizon. Even so, it is possible to see stars and other bodies situated up to around 50° south of the celestial equator (see Figures 3.15 and 3.16).

We can completely see, for example, the constellation of Scorpius, entirely situated in the southern celestial hemisphere. This is visible from latitudes near 40° N, on summer nights, on the southern horizon. From March onwards it is also possible to see Scorpius in the south just before sunrise.

Moving back north, looking at, say, the constellations of Ursa Major and Cassiopeia, one notices that their stars rotate approximately around Polaris, anti-clockwise (see Figure 3.7). The arrows indicate the direction in which the sky rotates. As this movement is slow, it is detected more easily when you compare the sky's appearance on two separate observations, for example one hour later, using a tree or a building as your reference point. You will also see that when you look north, the stars rise on your right, make arcs of different sizes, and set on your left. Such considerations, as in Figures 3.7 and 3.10, refer to the sky observed from latitudes close to 40° N, such as the central belt of North America or mainland Europe. In Chapter 5, or more specifically in Section 5.3, the reader will find references applicable to locations in the southern hemisphere.

Looking east, the stars rise obliquely on the eastern horizon (see Figure 3.8) and set in the west, also at an angle to the horizon. Looking south the stars move from left to right in virtually parallel lines to the horizon (see Figure 3.10). The angle shown in Figures 3.8 and 3.9 depend on the latitude of our viewing location, as shown in Figure 3.16.

We see that in all cases referred to above, the stars "move" along arcs parallel to the celestial equator. All arcs have their centre at the celestial poles. All arcs seen in the northern hemisphere are described as having the *celestial north pole* as their centre. This coincides approximately with the position of Polaris, as described above. We know that this is all a consequence of the Earth's rotation and that this "movement of the stars" around our planet is pure illusion.

Given that the sky appears to rotate *around* the celestial poles, the celestial pole that can be seen (be it north or south depending on where you are) is at a constant altitude above the horizon. Known as the *polar altitude*, this altitude is equivalent to the latitude of the place in which the observer is located. For example, if the reader lives in Boston, Massachusetts (latitude 42.4° N), the celestial north pole is at an altitude of 42.4°, on the northern side of the horizon. In Richmond, Virginia, on the other hand, the pole will be at a lower altitude (37.8°). Table 3.1 shows approximate latitudes of the main American cities.

To an observer standing at the Cape of Good Hope, latitude 34° S approximately, the celestial south pole will be at a altitude of 34° on the southern horizon (the celestial north pole will be inaccessible from this location, as it is below the horizon). This is described in Figure 5.1.

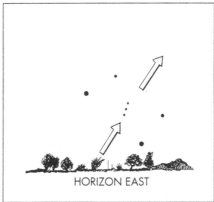

Figure 3.7. Apparent northward movement of the stars.

Figure 3.8. Apparent eastward movement of the stars.

Taking one particular star as a reference, observation will reveal that the star rises each day 4 minutes *earlier* than the day before (3 min, 55.9 sec, to be precise). This detail will be explained later in the book. It is because of this small difference, repeated every day, that the sky's appearance, observed at the same time on consecutive days, is slightly different each time. This difference of almost four minutes a day

Figure 3.9. Apparent westward movement of the stars.

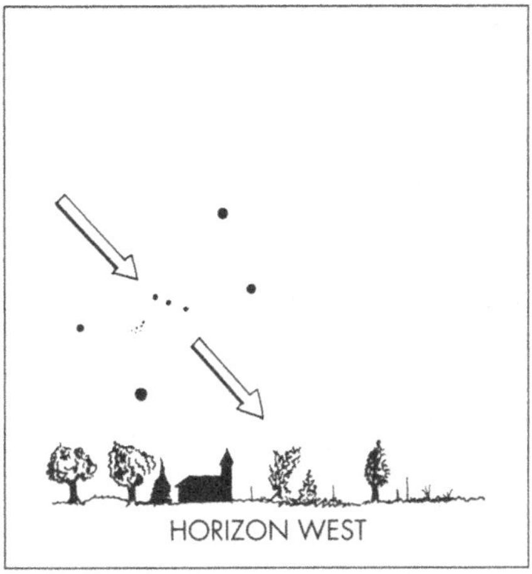

Figure 3.10. Apparent southward movement of the stars.

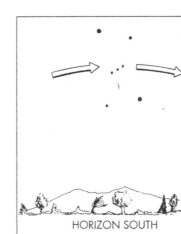

HORIZON WEST

HORIZON SOUTH

Table 3.1. Approximate latitudes of some American cities

City	Latitude	City	Latitude
Austin	30.4°	Minneapolis	44.5°
Boston	42.4°	New Orleans	30.0°
Chicago	41.7°	New York	40.8°
Cincinnatti	39.2°	Philadelphia	40.0°
Dallas	32.5°	Pittsburgh	40.8°
Denver	39.7°	Salt Lake City	40.7°
Kansas City	39.0°	San Francisco	37.8°
Los Angeles	33.6°	Seattle	47.3°
Memphis	35.5°	St Louis	38.5°
Miami	25.5°	Washington DC	38.9°

accumulates: at the end of a month, it reaches, approximately: 4 minutes × 30 days = 120 minutes = 2 hours.

Since the celestial sphere does one complete 360° rotation in around 24 hours (23 h, 56 m, 4.1 sec., to be precise), it therefore rotates at around 15° per hour (given that 360° / 24 hours = 15° per hour), or 30° in the two hours mentioned above. So if you look at the constellations at the same time each day, they will move 30° to the west in one month. In other words, the sky's appearance when viewed from one particular location, on, say, 1 January at 11 p.m. will be the same as on 1 February at 9 p.m. or 1 March at 7 p.m.

When it is still dark enough before the Sun has risen, identify stars in the east and you will see that, week on week at the same time, the same stars will come up higher and higher in relation to the horizon (see Figures 3.11 and 3.12). Gradually other stars, previously hidden by sunlight, will appear on the horizon. The sky's appearance, observed at the same time of day, is therefore characterised by the time of year. At the end of the year, the same constellations begin once again to rise at the same time in relation to the Sun.

For the same reason, if you identify stars visible in the west, following sunset, just as the sky is beginning to darken, the same stars will start to come up lower and lower in relation to the horizon (see Figures 3.13 and 3.14). They are eventually hidden from view, lost in the twilight. Others will appear to take their place, followed by further stars, and so forth. At the end of the year, you start to see the same stars again after sunset. I will return to these subjects in more detail in the section on the ecliptic.

Some stars, seen from a particular location, never set. They are always above the horizon while they circle around the celestial pole visible from the location. This is not a characteristic of these stars, of course, but simply a consequence of the Earth's rotation and of the observer's geographical position. Such stars rotate around the celestial pole visible from the location above the horizon in question. However, as they are especially close to the celestial pole, they remain above the horizon as they rotate. They are therefore referred to as *circumpolar stars*. Figures 3.15 and 3.16 illustrate this and show what is required for a star, seen from a particular location, to be considered circumpolar.

In the central belt of the USA, for example, the stars of Ursa Minor and Cassiopeia and most of Ursa Major, among others, rotate around the celestial north pole without ever dipping below the horizon (see Figure 1.8 and Identification Map M1).

The central belt of the USA extends approximately between 37° N and 42° N, as can be seen in Table 3.1.

The latitudes of some cities in the southern hemisphere, or close to the equator, in English-speaking countries, can be found in Appendix 7.

The central belt of the USA is at around 40° N. Observers in this region can take this 40° figure as the average latitude. At this latitude, all stars less than 40° from the celestial north pole are circumpolar (see Figure 3.16).

There is no great difference in the sky viewed from Boston or from Richmond. The observer in Richmond, Virginia can see areas of the celestial sphere located at around 5° further south than the observer in Boston is able to. This difference corresponds to the angular measurement made by the width of three fingers at arm's length. For the same reason, when observed from Boston, Polaris is 5° higher than when seen from Richmond.

There are other stars which cannot be seen from the central belt of the USA, as is the case of the Southern Cross, Pavo the Peacock and the Southern Triangle, among others. At 40° north, stars that are less than 40° from the celestial south pole can never be seen. This celestial pole is below the horizon for any observer in the Earth's northern hemisphere (see Figure 3.16).

For all locations on Earth (except places close to the equator) there are circumpolar stars and those that can never be seen. The limits corresponding to these two

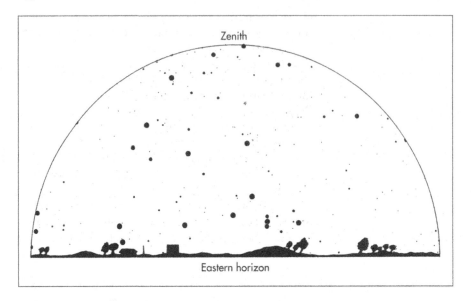

Zenith

Eastern horizon

Figure 3.11. Appearance of the sky on 8 August, in the east, at 5 a.m., official time (the time shown on normal clocks and watches). Compare this with the following diagram. The date could be in any year.

Figure 3.12. Appearance of the sky on 23 August, in the east, at 5 a.m., official time. The stars are higher in relation to the horizon than in the previous diagram. Notice that stars appear above the horizon that were not visible in the previous diagram. The date could be in any year.

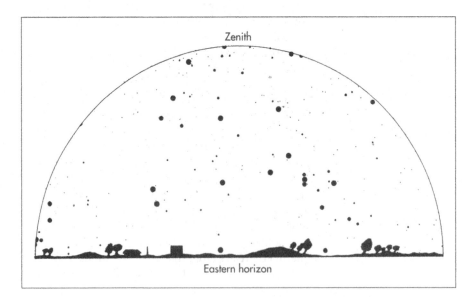

Zenith

Eastern horizon

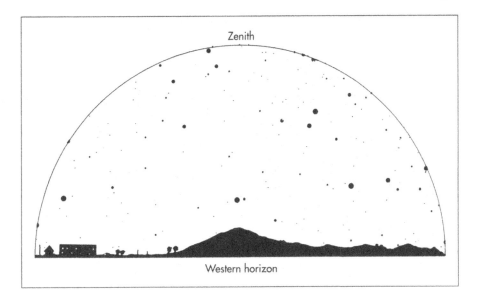

Figure 3.13. The sky's appearance on 10 December, in the west, at 7 p.m., official time. Compare this to Figure 3.14. The date could be in any year.

Figure 3.14. The sky on 30 December, in the west, at 7 p.m., official time. The stars are lower in relation to the horizon than in the previous diagram. Notice that stars that were visible just above the horizon in the previous diagram have now disappeared from view. The date could be in any year.

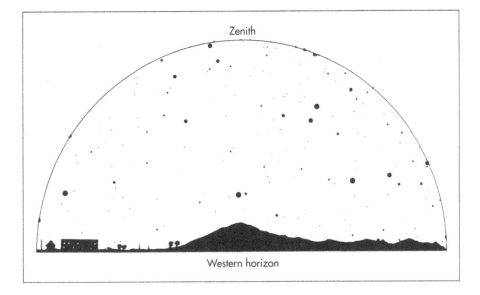

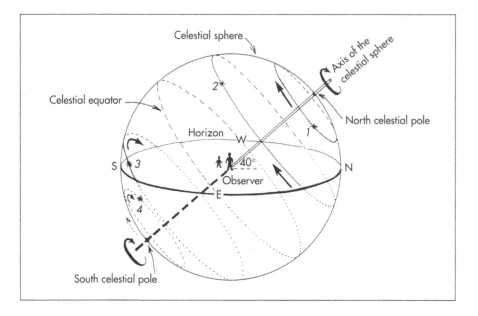

extremes depend on the latitude of the viewing location.

There are two special and fascinating cases. To people living on the equator, all stars rise and set. There are no circumpolar stars. The celestial poles are on the horizon, the celestial north pole in a northerly direction and the celestial south pole in a southerly direction. To people on one of the Earth's poles, no star rises or sets. All stars visible at any given moment are circumpolar (the corresponding celestial pole is at the zenith, the other being at the nadir). These two extreme situations can be interpreted easily, making a similar pattern to Figure 3.15, but adapted to the corresponding latitude (90° for the north pole and 0° for the equator). For the southern hemisphere, see Figure 5.1.

Figure 3.15. Stars that never set – circumpolar – and stars that never rise, viewed from a location at latitude 40° N. The observer is in the middle. Star number 1 is circumpolar; 2 and 3 rise and set, although 3 appears only just above the horizon and can only be seen looking south. Star number 4 does not rise (in the location in which the observation is taking place), as it will always be below the horizon.

3.1 So Just What Are The Constellations?

Nowadays, the constellations are defined as *regions of the celestial sphere*, marked out by borders defined in 1928 by the International Astronomical Union, in order to suit the shapes attributed to them by the

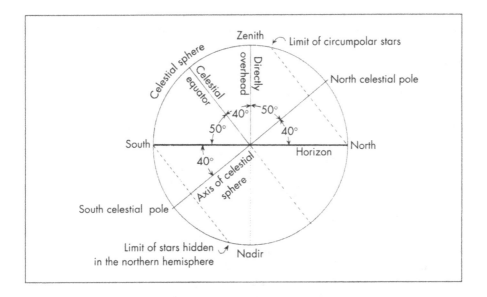

Figure 3.16. Situation in which certain stars can be considered circumpolar, and in which others do not rise, for someone observing at a location with a 40° N latitude. This latitude corresponds to the central belt of the USA, e.g. Philadelphia. You can see the angle made by the equator with the horizon. Note that in locations situated in the Earth's northern hemisphere, as in the USA's case, the celestial equator reaches its maximum height on the south side of the horizon. Using this diagram as reference, it is possible to make similar conclusions about locations with different latitudes, by replacing 40° with the relevant figure, changing other angles accordingly.

ancient world. The Union abolished some constellations and recognised 88 constellations altogether, covering the entire firmament. In this way, any point in the celestial sphere belongs to one constellation, and one only. It is inaccurate to say "there are 88 constellations", as this number could have been 50 or 300 or any number you could think of. The 88 constellations were the result an international convention. The names of all of the constellations are shown in Appendix 6, which also includes further information regarding visibility and location.

As mentioned in previous pages, our ancestors' association of the names of legendary figures with the configurations drawn by the stars in the sky was apparently a somewhat naïve process. It transformed a "jumble" of points into something coherent and easy to follow. It was a serious and methodical piece of work, so much so that the same *names* are still in use today, in accordance with tradition and a similar methodology.

Of course, nowadays we do not try to see an eagle "flying" in the night sky, in the position defined by the stars in the constellation Aquila (the Eagle), except for the purposes of historical curiosity. Nevertheless, any star, planet, nebula, galaxy, etc., which is known or which may become known *within* the borders set out for the constellation Aquila is considered, by the current astronomical community, to be in the constellation Aquila and is catalogued accordingly. This only means that the "object" in question is seen in that direction and does not mean in any way that it is related to the other stars in the constellation. "In the constellation of Aquila" automatically means "visible in the direction of the constellation of Aquila". Naturally this applies to the other constellations. The word "object" is used to designate any astronomical object, which can be a planet, a nebula, a galaxy, or whatever.

So, *Deneb* is the brightest star in Cygnus (the Swan) and *Antares* is the boldest in Scorpius. In the same way, we can say that, on a particular occasion, the planet Mars is in the constellation Gemini or that Saturn is in Capricorn. If a planet is seen on one day in the constellation of Taurus, it may be seen months later in Leo. For the purposes of astronomy, the word "is" means "visible in the direction of". It is worth bearing in mind that the borders of the various constellations are dictated entirely by the direction of observation. These borders are conventionally accepted in the same way as the borders that mark out the countries of the Earth. Some examples are given in Figure 3.17.

In some cases, so as not to repeat the word constellation, only the corresponding name will be mentioned. So when I say Taurus, I am referring to the constellation Taurus.

The stars from different constellations are not related to one another in any way. They only appear to be close together because they are aligned in the same direction when observed from Earth (see Figure 3.2). In fact, some may be hundreds of times closer to us than others. A constellation is therefore not an individual group of stars in space and it does not exist in this way. Figure 3.18 shows, for example, the different distances of the brightest stars in Orion from us.

You would only need to be a few dozen light years from the Sun, living on a planet orbiting another star,

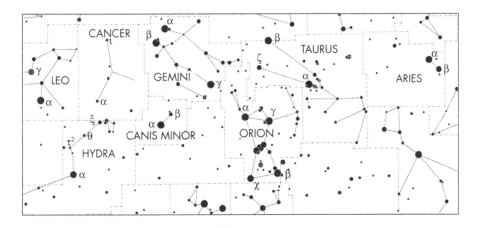

Figure 3.17.
Examples of
constellation borders.
The names of some of
the constellations
covered by the
diagram are indicated.
Compare this with
Figures 1.1 and 1.3,
which show the same
region of the celestial
sphere, but without
constellation groupings
or border markings.

for the panorama of constellations to be different from the one with which we are familiar on Earth. The constellations would form a different "shape" and, amazing as it may seem, some stars that we usually consider as belonging to one particular constellation could be seen in other constellations. In Figure 3.2, for example, stars A and B, viewed by an observer standing close to star C, would no longer be side by side in the sky.

When you contemplate the starry sky you have the impression that all is peaceful and in good order. This misleading appearance led people to think for many years that the Universe was always the same, working like clockwork without anything major ever happening. The extraordinary progress of astronomy and astrophysics in the last hundred or so years has shown this assumption to be false. The Universe evolves. The apparently tranquil night sky "hides" remarkable and intriguing phenomena and happenings. Some are rarer than others and most are not visible to the naked eye. Some even escape the view of telescopes and require sophisticated processes in order to be detected.

Just to give you a few examples, the Moon is moving slowly further away from the Earth. The planets of the Solar System are not as inert as previously thought. Saturn is not the only planet with rings. In July 1994, a comet crashed onto Jupiter. Stars do not live forever; they are born, evolve and die. Towards the end of their lives, large-mass stars explode, shining like a million suns. They leave behind rapidly expanding gas clouds and small, highly concentrated stars, which spin on their axes many times per second and emit powerful beams of radiation. One cubic centimetre of these

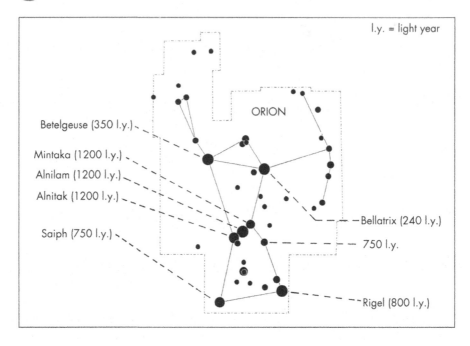

l.y. = light year

ORION

Betelgeuse (350 l.y.)

Mintaka (1200 l.y.)

Alnilam (1200 l.y.)

Alnitak (1200 l.y.)

Saiph (750 l.y.)

Bellatrix (240 l.y.)

750 l.y.

Rigel (800 l.y.)

stars' matter weighs thousands of tons. Those of even greater mass end their lives by giving off phenomenal concentrations of matter, which "clean" the space around and prevent light from escaping. Some galaxies have collided and, in a gigantic process, have reshaped each other. The Universe is expanding, as other galaxies move away from ours, at fantastic speeds proportional to the distance from us. There are also mysteries and questions that have yet to be cleared up.

Figure 3.18. Distance from Earth of the brightest stars in the constellation of Orion. The distances are shown in light years. The names of the stars are also indicated.

These extremely interesting happenings are outside the remit of *Navigating the Night Sky*. In fact, as mentioned in the introduction, the aim of this book is to familiarise the reader's naked eye with the constellations and their brightest stars, and for the reader to be able to interpret the different aspects of the sky according to time, date and viewing location. Knowledge of the night sky is also important for readers who wish to go on and discover where to point binoculars or a telescope. For example, if you want to look at the Andromeda Galaxy, located in the constellation of the same name (see Figure 4.8), you *need* to be able to identify the constellation and to know if it is going to be visible that night. You will also need to be able to locate the galaxy in relation to the constellation's stars, some of which will also have to be identified. The same will be true when you go on to observe other targets in various constellations that are accessible if you use binoculars or telescopes. Gradually it will be worth owning an observation instrument, if you wish to know more about the constellations.

There are two main reasons for bringing this to the reader's attention. Firstly, to remind the reader that the sky is not calm and peaceful, as it may superficially appear, and secondly, to show that it is essential to know the constellations and to be able to identify the brightest stars easily and confidently, even for those who wish to go beyond the study of astronomy.

In order to identify a constellation, you have to start by locating the brightest star, or two or three of the brightest and identify the other stars later.

In maps it is common to link the brightest stars by means of lines to make them easier to recognise. Please note that these are not the only stars in the constellation. In fact if you use binoculars, it becomes clear that the number of stars visible runs into the hundreds or even thousands. The use of a telescope will raise this number still further. So it is inaccurate to say "Ursa Major has seven stars". All stars and other bodies visible within a constellation's borders belong to that constellation, as mentioned above (see Appendix 1). In an excellent viewing location, on a clear night and assuming you have good eyesight, you will be able to see 120 stars in Ursa Major, with the naked eye. With binoculars you can find hundreds of stars in the constellation, and thousands if you use a telescope.

3.2 Measuring the Sky

Some stars seem close to others, while others seem further away. It is often useful for a sky explorer to know how to express and interpret such apparent distances between stars.

Let us take a simple, everyday example. Stand a few metres away from a person who is 1.70 m tall. Stretch out your arm and open your hand, making a vertical span, and you can see if the person is *apparently* bigger or smaller than your span.

If the person is 4.4 m away, your span will be "the same height" as the person. One of the tips of the span "touches" the top of the person's head and the other the bottom of their foot (see Figure 3.19). The angle made by the person's height in relation to your eyes is

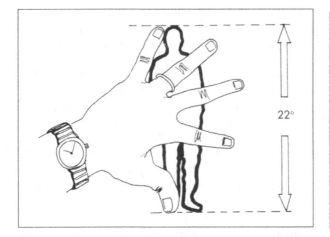

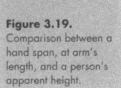

Figure 3.19. Comparison between a hand span, at arm's length, and a person's apparent height.

equal to the angle made by your hand span (see Figure 3.20). This measurement is only valid if your arm is outstretched. It is also necessary to use one eye only, keeping the other closed. If the person is shorter than 1.70 m then they will look smaller than your hand span at a distance of 4.4 m. Naturally, they will also look smaller if further away. If the person is taller than 1.70 m or if they are closer than 4.4 m they will appear to be bigger than your span.

At arm's length, your span makes an angle of around 22°. This means that the angle with the vertex in your eye, made by two lines, one along the top of your span and the other along the bottom, is 22° (Figure 3.20). Therefore, the *apparent* height of a 1.70 m-tall person 4.4 m away from your eye is 22°.

The width of the index finger (still at the end of your outstretched arm) makes an angle of around 2°. In the same conditions, a closed fist makes a 10° angle.

Figure 3.20. Angle of 22° at arm's length.

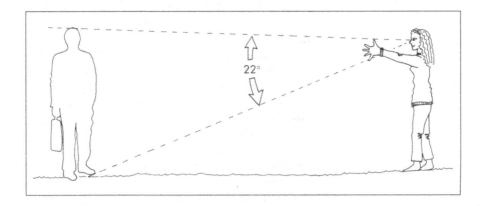

But, people's hands are different sizes, you might argue. Of course this is true, but people with small hands tend to have shorter arms and vice versa. For this reason, the process of measuring angles, while not absolutely accurate, is adequate for approximate measurement of angular distances between stars, between planets and stars, etc. Figure 3.21 shows some examples of this means of measuring angular distances.

This method also makes it possible to total up the results of measurements, using both the hand and the width of the fingers. For example, using both hands, one fist + one span = 32°; two spans = 44°; one fist + three fingers = 15°.

If you see two stars in the sky and, at arm's length, you measure that they are one span away from each other, you can say that the *angular* distance between the two is around 22° (angular distance is also known as apparent distance).

Of course this has nothing to do with the actual distance between stars, nor does it suggest that they are the same distance from us. It is simply that a line from one of the stars in the direction of your eye and a line from the other towards your eyes make an angle of 22°. If the apparent distance between them corresponds to the width of an index finger, one can say that they are approximately 2° apart, and so forth.

Using these processes for measuring angular distances it is now easy to see that, for example, the angular distance between the Pointers of Ursa Major (*Dubhe* and *Merak*) is around 5°. One can also understand what it means when we say, for example, that Dubhe (in the constellation Ursa Major) is around 28° from Polaris (in the constellation Ursa Minor). Figures 3.22 and 3.23 show a number of examples which will help the reader to get some practice. You should also try some of these out on "real" sky.

You can also use this method to measure the approximate height at which a star is located at a particular moment. Count the number of fingers that "fit" between the (unencumbered) horizon and the star. You can also measure the height of a planet in the sky, the angular distance between a planet and a star, or between a planet and the Moon at a given point in time. The *apparent* diameter of the full moon is around 0.5°.

It is not absolutely essential to know these techniques, but they are very helpful. It only takes a few days to know how to measure. Then you will be able to

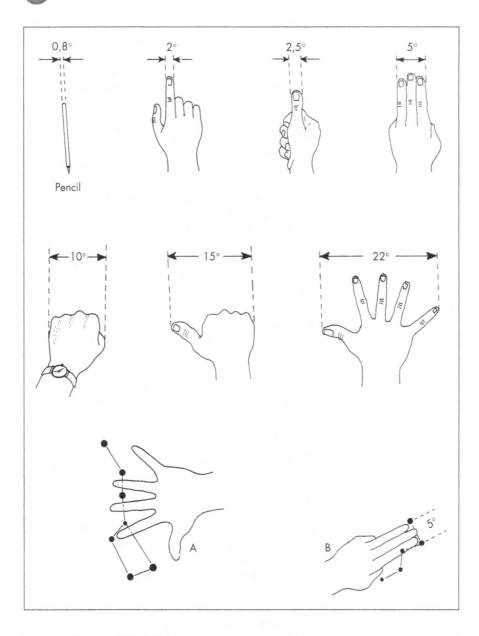

Figure 3.21. Method for approximate measurement of angular distances, using the hands and fingers. One can use the left hand or the right, whichever is more comfortable – it makes no difference. The arm must be outstretched for the measurements to be correct. (A), near the bottom, makes the comparison between the apparent (angular) size of a hand span and the size of the shape outlined by the seven brightest stars of Ursa Major. (B) measures the approximate angular distance between the two brightest stars in Cassiopeia, 5°. These measurements can only be made with one eye, keeping the other closed.

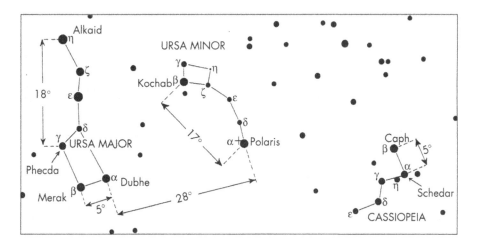

Figure 3.22.
Examples of angular distances between stars near the celestial north pole, which has already been identified in previous diagrams. The angular distance between Alkaid and Merak is around 25°. The + sign alongside Polaris shows the position of the celestial north pole.

say, for example, "last night, Jupiter was 4° west of the Moon". You can also show that, *in relation to the stars,* the Moon moves 13° to the east every 24 hours.

3.3 What Is the Ecliptic?

The scene is a living room. The reader, representing the Earth, moves slowly around a lamp, representing the Sun (see Figure 3.24). When you are in position 1, the lamp is in line with the chair (A); once you arrive in position 2, the Sun will be in line with the table (B); when you are in position 3, the lamp will be in line with the picture (C) on the wall. So as you orbit the lamp (which may be on or off for this experiment), it appears to move to the left in relation to the lounge background. You will only know you have completed one lap of the lamp when it is back in front of the same object in the lounge.

The Earth orbits the Sun, as the reader "orbited" the lamp in the diagram above. In the same way, at any point of the year, the Sun is aligned with a particular constellation and seems to move an average of 1° a day, *from west to east, in relation to the stars.* Consequently, during one year, the Sun goes on a trip around the celestial sphere, aligning with various constellations on its way round. These are the constellations of the zodiac (see Figures 3.25 and 3.26).

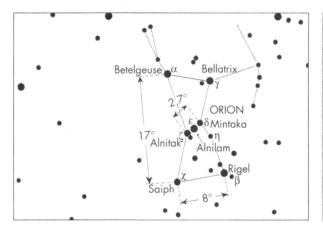

Figure 3.23.
Examples of angular distances between stars in the constellation Orion, one of the easiest to recognise. It is visible from December to April from nightfall, and from August to December at daybreak. During the month of December it rises at nightfall and sets at daybreak.

The route taken by the Sun in its annual trip is the ecliptic. Of course, this "movement of the Sun" is merely a consequence of the movement of the Earth around our star. This makes us believe that it is the Sun that is moving among the constellations, which are effectively the Sun's background. The zodiac is no more than a strip of the celestial sphere, centred on the ecliptic, which is 16° wide (8° to the north of the ecliptic and 8° to the south), as shown in Figure 3.26.

Naturally, given the brightness of the Sun, it is not usually possible to see it at the same time as the stars. But they are there, even during the day, and the Sun is in line with some stars, as shown in Figure 3.24, in which the lamp was in line with the table at a certain

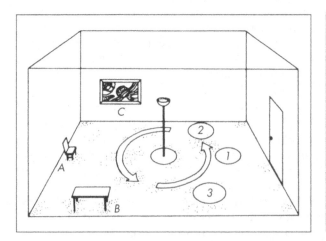

Figure 3.24.
Apparent movement of a lamp in relation to a background scene when an observer moves around the lamp.

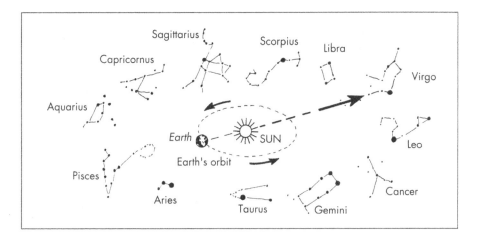

Sagittarius

Scorpius

Libra

Capricornus

Virgo

Aquarius

Earth SUN

Earth's orbit

Leo

Pisces

Aries

Cancer

Taurus

Gemini

point in time. On occasions when there is a total eclipse of the Sun, the brightest stars can be seen in the background, as the Moon has come in to block some of the brightness of our star.

The ecliptic can be seen differently, as in previous examples. If the reader considers the plane of the Earth's orbit stretched out into space, the ecliptic can be understood as the intersection of this plane with the celestial sphere (Figure 3.26).

Taking a certain constellation of the zodiac as a reference, one year later our star passes in the direction of the same constellation again, meaning that the Earth has completed one lap around the Sun (one year has passed). This all happens in much the same way as in the example of the lamp above.

Table 3.2 shows the approximate time of year in which the Sun passes in the direction of each of the constellations of the zodiac. These constellations, and others in the vicinity, cannot, therefore, be observed at the times of year shown. There is a slight difference in these dates from one year to the next, due to leap years. The discrepancy is never more than one day away from the date shown. The Sun passes by on the dates indicated or, at most, one day before.

Since the Sun takes one year to move around the ecliptic, returning to the same stars one year later, it completes one lap (360°) in one year, approximately 365.25 days. This corresponds, on average, to 360° divided by 365.25 days, which makes 0.985 degrees per day. Rounding up this figure we can say that the Sun moves around the ecliptic from west to east, at a rate of approximately 1° a day.

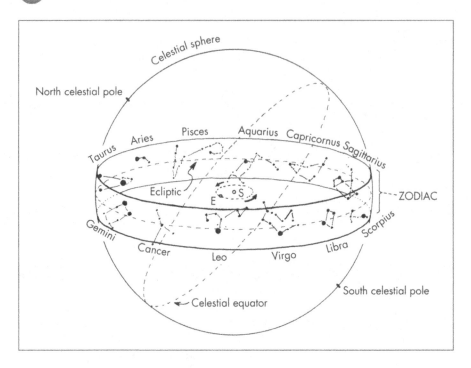

Figure 3.26. The celestial sphere, the celestial equator, the ecliptic and the zodiac. The constellations in the foreground are represented "in reverse". Letters S and E are the Sun (S) and the Earth (E). In this diagram the Sun is aligned with Capricorn. The diagram is not drawn to scale, of course, which would be impossible.

It is not possible for us on Earth to see the Sun and the stars at the same time. We can, however, verify that the constellations visible to the west, immediately after the Sun has gone down and darkness has fallen, change slowly as the weeks go by. They are not the same throughout the year; they *characterise* the time of year.

As the Sun moves through the ecliptic from west to east (in relation to the stars), between sunset on one day and sunset on the following day, a slightly *greater* amount of time passes than the time necessary for stars to become visible again in the same directions. Indeed, when the stars are visible again in the same directions, the Earth has turned once on its axis. But as the Sun has moved on the ecliptic about 1° to the east, the Earth, which rotates on its axis from west to east, will need to rotate a little further for the Sun to be seen in approximately the same direction in which it was seen the previous day. What is the difference between these two times? It is the time needed for the Earth to rotate on its axis approximately 1°, which takes around 4 minutes. This is why a particular star used as a reference rises every day around 4 minutes *earlier* than the day before, as mentioned previously.

In this way, the Earth rotates on its axis, in relation to the stars, in less time than it takes to complete one

Table 3.2. Position of the Sun in relation to the constellations of the zodiac

Approximate time of year	The Sun in line with the constellation
13 March to 19 April	Pisces
20 April to 14 May	Aries
15 May to 22 June	Taurus
23 June to 21 July	Gemini
22 July to 11 August	Cancer
12 August to 17 September	Leo
18 September to 31 October	Virgo
1 November to 24 November	Libra
25 November to 30 November	Scorpius
1 December to 18 December	Ophiuchus (see page 64)
19 December to 20 January	Sagittarius
21 January to 16 February	Capricornus
17 February to 12 March	Aquarius

turn on its axis, in relation to the Sun. Each star takes 23 hours, 56 minutes and 4.1 seconds to complete one apparent lap around the Earth, at the end of which it is once again visible in the same position. We can think of this as the time it takes for the celestial sphere to complete a lap around us on Earth. The Sun takes on average four more minutes than the stars to complete one (apparent) lap around the Earth. The slight difference between these times owes itself to the movement of the Earth around our star.

From week to week, the observer sees that the same constellations are increasingly lower on the horizon, in the west, until they disappear from view (see Figures 3.13 and 3.14). For each month that passes, the celestial sphere, viewed at the same time of night, appears to have rotated 30° to the west. Month after month, this alteration accumulates. Other constellations to the east take their place in relation to the Sun, and so the process goes on.

The same conclusion is reached, observing the constellations that are visible in the east, just before the Sun rises. As the weeks go by, the same constellations rise higher in the sky (see Figures 3.11 and 3.12). Other constellations appear, which were previously obscured by the Sun.

It has already been mentioned that the celestial sphere rotates at 15° per hour, from east to west (360° in around 24 hours). The gradual change in the sky's appearance, observed on successive nights at the same time, translates, as seen in the two previous paragraphs, to a monthly 30° alteration, which corresponds, in turn, to the rotation of the celestial sphere in two hours. Therefore, the sky's appearance on any given night, at, say, 10 p.m., is identical to what can be seen one month later at 8 p.m., and identical to what could be seen a month before at about midnight. This disposition of the stars in the sky is also identical to two months earlier at 2 a.m. or three months earlier at 4 a.m.

These observations lead us to believe that the Sun moves in relation to the constellations (a highly convincing illusion, caused by the movement of the Earth around the Sun). Taking one particular constellation as a reference point, the conclusion follows that it will be visible again, in similar conditions, one year later.

As shown in Table 3.2, the Sun, after passing through Scorpius and before it comes into line with Sagittarius, passes in front of a constellation called *Ophiuchus* (the serpent holder), which represented a man with a serpent coiled around his belt. From the point of view of astronomy, the Sun passes through 13 constellations in one year. Appendix 3 explains this in more detail.

Figures 3.27 and 3.28 offer the reader an overview of the ecliptic. It is indicated here by a broken line, giving a perspective of the ecliptic in relation to the constellations. The diagrams show north at the top and west on the right. Notice that there are no "waves" in the ecliptic. Its appearance in these two diagrams is a consequence of the scaling required to represent the celestial sphere on a flat surface, i.e. the page of the book. At the bottom, you can see dates corresponding to the various points through which the Sun passes. The reader can also see the celestial equator, whose plane makes an angle of around 23.5° with the plane of the ecliptic. The dates shown at the bottom make it possible to locate the Sun on the ecliptic and to conclude that this part of the sky and its vicinity (less than 20° east and the same to the west) is not observable on this occasion due to the sunshine. Those stars that, on any given occasion, are located to the left of the Sun (a little more than 20°) set a short time after the Sun on this date. The gap between the lines of two consecutive dates corresponds to around 15° in the sky, which enables us to make estimates of angular distances.

Figures 3.27 and 3.28 also show the approximate position of the Sun in relation to the constellations of the zodiac, on any date at any time, and are valid for

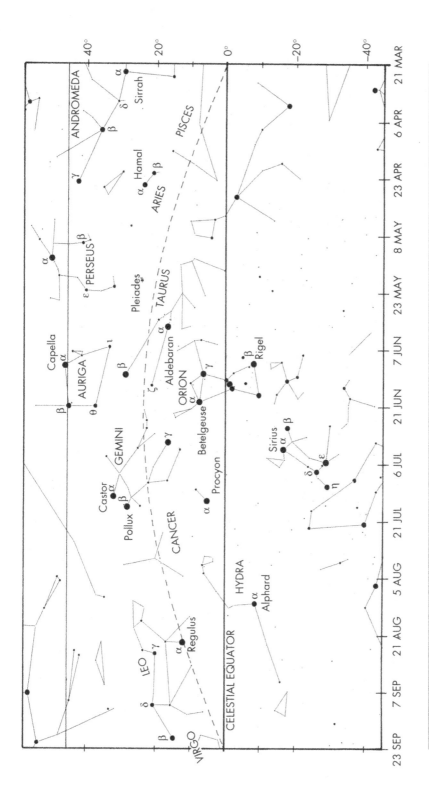

Figure 3.27. Representation of the ecliptic – broken line – from Pisces to Leo.

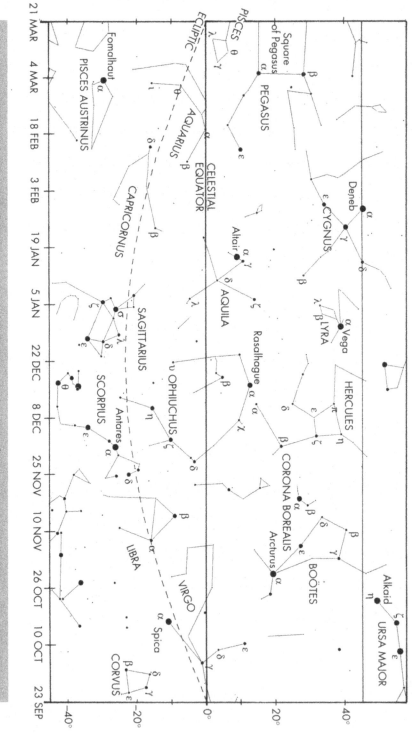

Figure 3.28. Representation of the ecliptic – broken line – from Virgo to Aquarius.

practically any year (at least during the next few hundred years). By way of an example, Figure 3.28 suggests that on 31 May of each year the Sun passes to the north of the star Aldebaran (α Tauri). On 6 July our star is aligned with the constellation of Gemini, so, on this date, as shown in the diagram, Aldebaran rises shortly before daybreak and the constellation Leo is visible in the west, just after nightfall.

These diagrams enable us to understand the notion of *declination*, which is very important in astronomy. The figures in degrees, on the right-hand side, are measurements in relation to the celestial equator, positive to the north and negative to the south. In order to make it easier to read these declinations, reference lines have been drawn on the left-hand side in the positions corresponding to the declinations shown on the right. The angle between one body and the celestial equator, perpendicular to the latter, is referred to as the declination of the body concerned. Thus the declination of any body situated on the celestial equator is 0°, on the celestial north pole 90° and on the celestial south pole −90°. Looking at Figure 3.28, the reader will see that the declination of the star Arcturus in the constellation Boötes is around 20° and that the declination of the star Spica in Virgo is approximately −11°. In one year, the declination of the Sun varies approximately between −23.5° at the beginning of the winter (in the Earth's northern hemisphere) and 23.5° at the beginning of the summer (see Figures 3.27 and 3.28). It is 0° at the start of the spring (approximately 21 March) and also at the beginning of autumn (around 23 September each year).

For periods of tens of years the declination of each star can be considered consistent (as far as naked eye observation is concerned).

3.4 The Ecliptic and the Positions of the Planets in the Sky

As the orbits of the planets are not especially inclined in relation to the plane of the Earth's orbit, the planets are always seen a few degrees from the ecliptic, to the north or to the south. This is why we *never* see, for

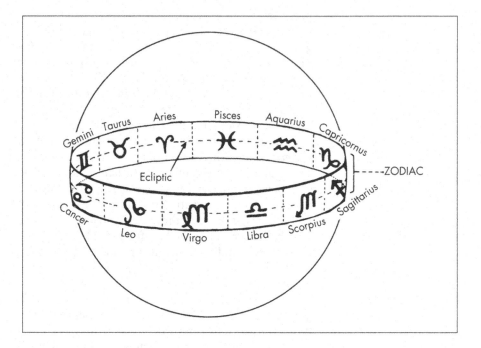

Figure 3.29. The zodiac and its signs.

example, a planet in the direction of the constellation Canis Major, Cassiopeia or Ursa Major. In other words, when the planets are visible it is because they are *close* to the ecliptic.

The zodiac is a strip of the celestial sphere, which includes the ecliptic and celestial regions between 8° to the north and 8° to the south (see Figure 3.29). Therefore the zodiac always contains the Sun, the Moon and the main planets, except Pluto. This occurs because the orbit of Pluto is more inclined than the others, in relation to the orbit made by our planet around the Sun.

Due to the movement of the Earth and the other planets around the Sun, the planets become aligned with different constellations. You can use the stars as a kind of backdrop, against which you can refer to the positions of the planets. Figures 3.27 and 3.28 give an overview of the ecliptic and are useful when mapping the possible positions of the planets.

Only five planets are bright enough to be seen with the naked eye. They are: Mercury, Venus, Mars, Jupiter and Saturn. Their luminosity is comparable to, or even greater than, a star of first magnitude. In order to identify a planet in the sky, you only need to know the constellations well or look at a night sky with one of the maps of the Celestial Chart in this book. Any bright dot that you see in the sky and which stands out from these maps is, in principle, a planet. If you are sure that "out there" in that region of the celestial sphere, there is no bright star, you can try to discover "who" it is that is standing out. Below are some tips to follow when you do not have a telescope at your disposal.

Mercury can only be seen shortly before sunrise, or just after nightfall. It is never more than 28° away from the Sun, since its orbit around the Sun stays "within" the Earth's orbit.

Venus is brighter than any star, but is never more than 48° away from the Sun. Therefore it can only be seen before sunrise, as "the morning star", or after nightfall as the "evening star". Neither of these two planets can be seen in the middle of the night. Mercury and Venus make smaller orbits around the Sun than the Earth does, and so they move fast. In just a few weeks you can see that their position in relation to the constellations has changed quickly. It is also worth bearing in mind that Venus is much brighter than Mercury.

Mars is highly distinctive and impossible to confuse, with its reddish glow.

Jupiter is, usually, the brightest planet, after Venus. But as the brightness of the planets depends on their positions in relation to the Earth and the Sun, whose light they reflect, Jupiter and Venus look, at first glance, similarly bright. If the observation is done at midnight, i.e. a good while after the Sun has gone down, the planet you see could never be Venus. If these planets are seen shortly before daybreak, or just after nightfall, you should use the following criterion: if it is Venus, you will see after some weeks that it has moved a fair distance away from its previous position in relation to the stars; whereas Jupiter will be almost exactly in line with the same stars.

Saturn is less bright than Jupiter and has a more yellowish colour. Its change of position in relation to the stars is even slower than Jupiter's.

These changes of position in relation to the stars are never so quick that you can spot them merely by keeping your eyes on the planets for a certain amount of time. If you see a bright dot moving in the sky, it could be a man-made satellite or an aeroplane reflecting the Sun's light, but never a planet.

When the atmosphere is reasonably calm, the planets do not twinkle if they are high above the horizon: their shine is constant, totally the opposite of what happens with the stars. On nights with an uneven atmosphere, planets may twinkle, especially if they are not far above the horizon. There are rare occasions when the atmosphere is extremely calm, in which the stars close to the zenith do not twinkle. The criteria for the planets not twinkling should be approached with caution and in conjunction with other rules, so that you can be absolutely sure.

Astronomers use symbols (see Appendix 10) to represent the Sun, the Moon and the planets in tables and in records of happenings. Astrologers use the same symbols in their horoscopes.

3.5 The Ecliptic and the Signs of the Zodiac

The ecliptic was first identified by the Babylonians, between the ninth and third centuries BC (it is not known exactly when). It was also the Babylonians who, some 2500 years ago, divided the zodiac into 12 equal parts, along the ecliptic. They gave each of these parts a name and a sign. Each sign of the zodiac took up 30° ($12 \times 30° = 360°$).

Each of the 12 signs acquired the name of the constellation in which it was located. For example, the sign that contained the constellation of Leo was called the sign of Leo and so on. Although the signs were all 30° along the ecliptic, the constellations, as we know, come in various sizes: Leo is much larger than Cancer, for example. The correspondence between signs and constellations is no more than approximate.

At *that* time, each sign approximately contained the constellation of the same name. When the Sun was in the sign of Taurus, for example, it was in line with the constellation in question. It is believed that the signs were linked to people's everyday lives, such as harvest and religious practices, performing a useful, practical function.

Nevertheless, due to the precession of the equinoxes, a phenomenon with which we shall not be dealing in *Navigating the Night Sky*, the correspondence between *dates* and *constellations of the zodiac*, indicated in Table 3.2 (see also Figures 3.27 and 3.28), changes gradually. A constellation takes around 26,000 years to return to the same dates. The dates shown in this table refer to the current era.

Table 3.3. Position of the Sun in relation to the signs of the zodiac

Time of year	Corresponding sign	Symbol of sign
21 March to 20 April	Aries	a
21 April to 21 May	Taurus	b
22 May to 21 June	Gemini	c
22 June to 23 July	Cancer	d
24 July to 23 August	Leo	e
24 August to 23 September	Virgo	f
24 September to 23 October	Libra	g
24 October to 22 November	Scorpio	h
23 November to 21 December	Sagittarius	i
22 December to 20 January	Capricornus	j
21 January to 19 February	Aquarius	k
20 February to 21 March	Pisces	l

The *signs* shown above are still used in *astrology* but do not *currently* refer to the dates shown in Table 3.2. They were established around 25 centuries ago, when they *actually corresponded* (albeit approximately) to constellations with the same name. However, astrologers never updated them.

A quick glance at the astrology page of any newspaper will show that the time of year corresponding to each sign does not tally with the dates mentioned in Table 3.2; neither do the dates during which the Sun passes in the direction of a particular constellation coincide with the dates referring to the sign of the same name, shown in Table 3.3. Thus, currently, when the Sun is in, say, the *sign* of Cancer, it is actually in the direction of the *constellation* of Gemini. When the Sun passes in the direction of the constellation of Taurus, the corresponding sign is Gemini, and so forth.

If you were born on 10 November, for example, you "are" a Scorpius, as indicated in Table 3.3. On the day you were born, however, the Sun was in the direction of the constellation of Libra (see Figure 3.28). If you were born on, say, 9 December, you "are" a Sagittarius, but on the day of your birth the Sun was in the direction of Ophiuchus.

At the moment, the astrological *signs* do not coincide with the *constellations* that share their names. There is currently a time discrepancy of one sign, which will gradually increase over the course of the centuries. It takes 2160 years for such a one-sign time lag to occur. For the reasons stated above, the sign in which each planet *is situated* on a particular date does not currently correspond to the constellation in whose direction it is in fact observed, on the same date. For example, if on a certain date Jupiter "is" in the sign of, say, Virgo, it will be observed on that date in the constellation of Leo. 2500 years ago, however, the two did coincide.

During a period of around 26,000 years, each sign "does the rounds" of all of the constellations and temporarily resumes its position corresponding to the constellation with which it shares its name. 2160 years multiplied by 12 equals approximately 26,000.

Astronomers use the symbols of the zodiac signs to show positions of reference in the celestial sphere and do not attach to them any other meaning, whereas astrologers use these symbols when making their horoscopes.

3.6 How to Find Your Way Around the Sky

It is important not to forget that the stars visible in the west, at nightfall, are to the *east* of the Sun. Similarly, the stars visible in the east, before daybreak, are to the *west* of the Sun. West is always the direction in which the Sun "moves" in its apparent *daily* movement around the Earth. The direction in which the stars "go" in their apparent daily movement is also westward, as the celestial sphere appears to rotate from east to west due to the fact that the Earth's rotation goes the other way.

At any point on the celestial sphere, a direction parallel to the celestial equator is going to be east–west. We look westwards if we go in the direction of the celestial sphere's apparent movement and vice versa. From any given star, the direction of a line (drawn "straight" in the sky) linking this star with Polaris is *always*, almost exactly, *northwards*, i.e. towards the celestial north pole. Naturally, if you follow the opposite path you will be travelling south.

When we refer to "a straight line drawn in the sky", such as between two stars, we always mean an imaginary arc in the celestial sphere. The perspective we have of the sky is a spherical representation. Indeed, a map capable of encompassing all the sky visible at a given point in time, without any distortions, would have to be hemispherical. The observer would be in the middle of such a map, so as to recreate the view of the sky. This is what happens in a planetarium, where constellations are projected onto a giant hemispherical screen and in which the visitors are fairly close to the centre.

The celestial equator rises from the eastern tip of the horizon, from left to right, reaching its maximum height in the south and, as it goes back downwards, it crosses the horizon again at the westernmost point. The remaining part of the celestial sphere is below the horizon. In the south, a small slice of the celestial equator would be horizontal. These indications refer to a viewing location in the northern hemisphere, such as North America or Britain.

Due the Earth's orbit, the Sun appears to move on the ecliptic from west to east, as mentioned above. The stars rise every day 3 minutes and 56 seconds earlier than the day before and are approximately 1° further west than they were the day before at the same time. This difference is 7° a week, 15° in two weeks and reaches 30° in a month. If a particular star rises, for example, at 11 p.m. on a particular night, it will rise at 10 p.m. two weeks later. The hour difference is because the celestial sphere "rotates" from east to west at 15° per hour. After one year at the same time the stars are seen again in the same directions, as the discrepancy has reached 360° (30° × 12 = 360°).

These simple calculations explain many apparently difficult interpretations and make it extraordinarily easy for the observer to get around in the sky. Please see also Figure 4.2.

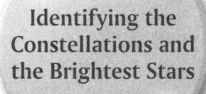

CHAPTER FOUR

Identifying the Constellations and the Brightest Stars

4.1 Alignments of Stars

Let us suppose that the reader already knows how to recognise some stars in the sky. When you look at them it is possible to imagine a straight line drawn in the sky linking one star to another. We say that these stars are in *alignment* in the sky, i.e. the two stars *mark out a certain direction in the celestial sphere*. Naturally, this line can be extended one way or another. If this alignment leads us to another star we say that this star is in the alignment defined by the other two (sometimes we follow an arc and not a straight line).

Let me show you an example, using probably the most famous alignment of stars, the one drawn from *Dubhe* and *Merak*, the Pointers of Ursa Major, which points towards Polaris, the brightest star in Ursa Minor. Since we have already identified the constellations of Ursa Major, Ursa Minor and Cassiopeia in previous diagrams, you should no longer have any difficulty in recognising them. Imagine a straight line drawn in the sky between Dubhe and Merak. Supposing we were to extend this line in the direction of Merak to Dubhe, we would eventually arrive at Polaris. Note that this extension is done in the opposite direction to the curve of Ursa Major's tail.

Alignments of stars are usually shown with arrows, some wider and some narrower, the wider ones being those which should be learnt first. The extended arrows take the reader's eye to other stars, which you will start to recognise if you follow this book's instructions.

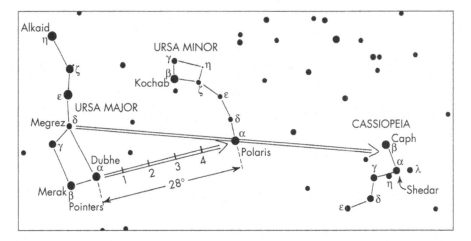

Figure 4.1. Examples of alignments. Starting point – the constellation of Ursa Major. Explanation of the steps to follow in order to locate Polaris and the constellations of Ursa Minor and Cassiopeia.

From now on we will be using alignments of stars systematically.

The Identification Maps are accompanied by explanatory notes about various alignments and regarding the stars and constellations that are next to be discovered. All you need to do if you wish to start to learn the sky is follow with your eyes the direction drawn by each alignment and compare the maps to what you see in the sky.

One way to reach Polaris from Dubhe is to calculate five times the distance between the Pointers, as shown in the diagram. This could be expressed as moving 28° in the sky (approximately, one span + three fingers at arm's length) from Dubhe to Polaris. In certain cases, the second of these recommendations is especially useful, as the angle in the sky is all you need. Using the hand method, with the outstretched arm, you get an idea of the angle between the star where the arrow begins and the star you wish to locate (where the arrow ends). Whichever you choose there is usually little chance of confusion, since the star you are looking for is, more often than not, the brightest in its region of the sky.

Let us take another example, also using Figure 4.1. Imagine a line from *Megrez*, also in Ursa Major, as far as Polaris (also known as the North Star). If you extend this line further, you eventually arrive at Cassiopeia, a constellation which looks something like a letter W when it is underneath Polaris; otherwise it resembles an M. The two brightest stars in Cassiopeia are *Shedar* and *Caph*.

There are other examples in which it is worthwhile making use of geometric figures. For example, certain very bright stars make a conspicuous triangle in the sky, or a square, and this makes it easier to recognise this group of stars and the constellations to which they belong. Notice that these stars can belong to different constellations. The *summer triangle* (see Maps M6 and C6) is almost isosceles, whose vertices are marked by the stars *Vega* (α Lyrae), *Deneb* (α Cygni) and *Altair* (α Aquilae).

By looking at the sky and the maps, and by following the suggested alignments you will find the stars and constellations shown.

I have already mentioned that the stars' alignments are shown by means of arrows. These work as a kind of rough map, featuring roads that enable us to travel from one place to another; as on a road map, there are "main roads" and "minor roads". If you follow them you will find observation much easier.

Taking "main roads" will lead you to the constellations that are easiest to identify and to the brightest stars. You should only take the "minor roads" when you want to find the less conspicuous constellations with fainter stars that are less easy to detect, constellations which seem only to "have" two or three stars, at first glance. The stars you are going to learn to identify, located by means of alignments, and the constellations you go on to recognise will both help to find other bright stars, and so forth. You can also try your own alignments.

In order not to repeat the word constellation too many times, from now on I shall only use the constellations' names. So, designations such as Gemini and Leo refer, of course, to the constellations of these names.

The positions of the planets are not indicated on the maps, as these positions change constantly. A celestial map showing the positions of the planets would only be valid for one month at the most and even in a few short weeks, one would notice that some of them have moved from the positions indicated.

The relative positions of the stars change so slowly that they can effectively be considered consistent for many generations. Maps that show stars and constellations are therefore valid for many years and can be thought of as perpetual. As far as naked eye observation is concerned, no significant alteration is expected in terms of the "shape" of the constellations.

4.2 Sky Maps and Road Maps

The similarity between identifying a place on a road map and identifying stars and constellations in the sky is greater than you might think. The reader's experience in reading maps will come in useful. When you are looking at the sky with this book in your hand, it is sometimes worth putting yourself in the shoes of a road-map reader. Let us have a look at some similarities and differences:

1. We have seen that we can use *alignments* to locate stars and constellations. These alignments are comparable to roads on a map, on which there are motorways linking major cities, as well as B-roads, smaller roads and byways linking smaller places. Similarly there are alignments in the sky that are easier to follow, which make it possible to locate brighter stars and the more conspicuous constellations. There are also "B-roads" allowing you to get to constellations characterised by fainter stars, making them more difficult to find.

2. Just as Rome is the capital of Italy, we can say that, for example, *Antares* is the brightest star in Scorpius. The reader will gradually become familiar with the brightest star in each constellation, where this is possible. All the 500,000-plus stars catalogued have designations, but only around 200 have their own names. In this book the names of around a hundred stars are mentioned.

3. If you ask someone the way from place *A* to place *B*, you will be told the roads to take, the main places you will pass along the way, the distance and any diversions that you may need. To find a star or a constellation, you can also follow the alignment made by two particular stars, go past other stars, turn right, etc.

4. If you want to find a small village on a map, you first need to know that it is near another, better known place. It is also useful to know if it is to the west or north of this place. It would be helpful to know that the village is, say, 30 miles from Huddersfield and not 150 miles. As for the sky, Identification Maps offer similar indications. You look at the stars with even greater confidence. You can find a certain star

even more easily if you know that it is, for instance, 10° west of another that you are familiar with, and not 50° south. The search is easier still if we know that the star is especially bright or not, if it is yellowish or reddish, etc. So, it is worth having a look at the map beforehand to check if the constellation you are looking for is north, south, east or west of another that we know. For example, Sagittarius is east of Scorpius, and Auriga is north of Orion, Canis Major and Canis Minor (see Map M3). This will avoid the pain of looking for the constellation in an area in which it is *not* located, which would be like looking for Liverpool in Humberside, or Newcastle in Hampshire. Knowing what is near what, and on what side, is as important in sky exploration as it is in reading the Earth's road maps.

5. Try not to be daunted by the volume of names in Appendix 4, just as the large number of place names on a map does not worry you. Just as with a road map, there is no obligation to know all of them by heart. You only need to be concerned with what you are trying to find on that particular occasion.

6. There are differences, of course, between sky maps and road maps. In the latter, places are in fixed locations on the surface of the Earth, whereas in the sky the stars are not – although they may *appear* to be – the same distance away from us. One can see "next to each other" stars that are at substantially different distances. Take the examples of Sirius and Wezen in Canis Major. Sirius is a near neighbour of ours, a mere 8.6 light years away. Wezen is more than 2000 light years away. These stars can be seen on Map M3 (see also Figure 3.2).

7. When you look at a map of the world, you see the Earth from the outside looking in. When you contemplate the stars, it feels as though you are in the middle of the celestial sphere, looking outwards to this immense sphere, which only "exists" in the imagination.

4.3 Using Identification Maps

These maps show the procedures to follow if you want to identify bright stars and virtually all the constellations in the night sky (constellations that are seen only in the Southern hemisphere and at low latitudes in the Northern hemisphere are in Map M8, Chapter 5). All of the maps are prefixed with a letter **M**. Each Identification Map contains a least one constellation or star that the reader already knows from the maps and diagrams on previous pages. These stars or constellations will be pointed out on each map as the *starting point*. Each Identification Map appears on the right-hand side of the page, and features the alignment arrows you will need to follow. On the left you will find various indications regarding the alignments or geometric figures that will be useful to help you identify constellations and the brightest stars in that particular area of the celestial sphere, as well as a few explanatory notes. The maps should actually be used when you look at the night sky, as this is the only way to make identifications.

So as not to have the arrows all the time, Chapter 7 contains a Celestial Chart, whose maps are prefixed with the letter **C**. Each of these shows stars and constellations covered by the Identification Map of the same number (Map C3 shows the same celestial area as Map M3). Each map of the Celestial Chart appears on the left-hand page, accompanied on the right by another map, usually without a key and in positive relief, in order to train the reader.

I have used dots of different sizes to give the reader an idea of the approximate brightness of each star; the brighter the star, the bigger the dot. There is a key at the top of each map, which will enable the reader to make an estimate of the brightness of each star. Although some stars appear brighter than others, they do not look bigger than each other.

In order to convey both the "size" of the dots and the angular distance between the stars, the dots sometimes seem to be almost on top of one another. Looking at the sky, one notices that the angular separation between the stars is abundantly clear, albeit relatively small. This happens, for example, with the three stars of Orion's belt (see Map M3 and Figure 4.9).

The Identification Maps and the maps of the Celestial Chart show all the stars that can be seen with the naked eye, from a reasonable viewing location, up to magnitude 4.8. I had to find a happy medium; maps that only showed the very bright stars would be of little use in real life, whereas a map showing too many stars, including those invisible to the naked eye would be too confusing in terms of finding the star that corresponds to the name in the key. The maps are decent representations of the sky seen from a reasonably dark location. From an excellent location you will see more stars than you will find on these maps, but these additional stars will be a lot less bright, and difficult to make out with the naked eye. Appendices 5 and 6 give useful indications regarding ease of identification, visibility and location of all constellations.

As it is impossible to represent the entire celestial sphere on one map, I needed to make several. This necessitated splitting up some constellations. In order to make it easier to connect each map, there is always overlap shown in neighbouring maps. The reader will therefore find some stars and constellations which repeat, from one map to the next. The continuation to the left (east) and to the right (west) of the part of the sky visible in each rectangular map is in the preceding and following maps. The part of the sky which lies to the right (west) of Map M7 can be seen on Map M2.

There are eight main maps, covering the whole of the celestial sphere. Of these, two are circular and encompass areas that are relatively close to the two celestial poles. The other six are rectangular and show the areas between the two circular maps. There is also some overlap between the circular maps and the adjacent parts of the rectangular maps.

The names of the months of the year make it easier to link the various maps.

Figure 4.2 shows how the celestial sphere has been divided in order to make these maps. The six rectangular maps correspond to the plan of the lateral surface, divided into six, and the two circular maps are the plans of the two "lids" in the vicinity of the celestial north and south poles.

At the bottom of each rectangular map you will find the names of the months of the year. Choosing the appropriate month for the time of observation, these maps show the stars passing in a southerly direction on the horizon at around 9 p.m. (10 p.m. during summer time). However, the maps can be used at any time of the night, since, as mentioned above, the sky's appearance at a particular time in a particular month (as far as the stars and constellations are concerned) is the same as one month later, two hours earlier. This is also the same as one month before, two hours later. So, the sky's appearance in mid-November at 9 p.m. is identical to what can be seen in mid-December at 7 p.m.

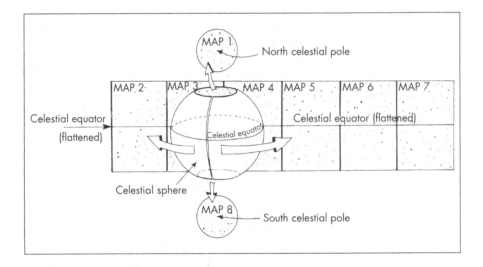

Figure 4.2. Mapping of the celestial sphere, leading to the eight Identification Maps and celestial plan maps. The overlap between adjacent maps is not represented here. Naturally, this mapping leads to some distortions of the aspect of the constellations. But dividing the celestial sphere into various sections means that the distortion is not very major; the shape of the constellations, as seen on the maps, is very similar to what you see in the sky and as such the distortion will not cause the reader much difficulty. In rectangular maps the top and bottom are the parts that have undergone the greatest reshaping. These areas are also at the edges of the circular maps, where they are shown with less distortion. Some smaller scale maps show certain regions of the celestial sphere in greater detail.

and in mid-October at 11 p.m. If you are observing at the times shown (9 p.m. winter time and 10 p.m. summer time) use the name of the actual month and choose the appropriate map looking south. If you are observing at 7 p.m., use the month after the month in which you are in fact observing, i.e. if it is December, use November. At 11 p.m., use the following month to the one you are in; at 1 a.m. two months later, and so on. If you are looking north always begin with Figure 4.5 and Map M1.

Once you have identified the constellations visible on the southern horizon, use these constellations to explore eastwards (left) or westwards (right), using the same maps. Following the indications given, you can also recognise constellations to the north (further up on the maps), right up to the zenith and also those to the north of the zenith.

Table 4.1 shows all possible permutations, the time of year, the time of day and the map which shows the constellations visible to the **south**, in this case. This table also helps you to orientate the circular maps (M1 and M8), and all of the maps of the Celestial Chart. Of course, maps can be dovetailed to the beginning, middle or end of each month.

In Table 4.1, the spaces marked with a "–" show that at these times the sky is not dark enough to allow the stars to be observed.

Table 4.1. Selection of Identification Maps and maps of the Celestial Chart, according to the date and time of observation

Month of observation	Time of observation	Zone of map for orientation	Month of observation	Time of observation	Zone of map for orientation
	7.00 p.m.	December		–	June
January	9.00 p.m.	January	July	10.00 p.m.	July
	11.00 p.m.	February		0.00 a.m.	August
	1.00 a.m.	March		2.00 a.m.	September
	3.00 a.m.	April		4.00 a.m.	October
	5.00 a.m.	May		6.00 a.m.	November
	7.00 p.m.	January		–	July
February	9.00 p.m.	February	August	10.00 p.m.	August
	11.00 p.m.	March		0.00 a.m.	September
	1.00 a.m.	April		2.00 a.m.	October
	3.00 a.m.	May		4.00 a.m.	November
	5.00 a.m.	June		6.00 a.m.	December
	7.00 p.m.	February		–	August
March	9.00 p.m.	March	September	10.00 p.m.	September
	11.00 p.m.	April		0.00 a.m.	October
	1.00 a.m.	May		2.00 a.m.	November
	3.00 a.m.	June		4.00 a.m.	December
	5.00 a.m.	July		6.00 a.m.	January
	8.00 p.m.	March		8.00 p.m.	September
April	10.00 p.m.	April	October	10.00 p.m.	October
	0.00 a.m.	May		0.00 a.m.	November
	2.00 a.m.	June		2.00 a.m.	December
	4.00 a.m.	July		4.00 a.m.	January
	6.00 a.m.	August		6.00 a.m.	February
	8.00 p.m.	April		7.00 p.m.	October
May	10.00 p.m.	May	November	9.00 p.m.	November
	0.00 a.m.	June		11.00 p.m.	December
	2.00 a.m.	July		1.00 a.m.	January
	4.00 a.m.	August		3.00 a.m.	February
	6.00 a.m.	September		5.00 a.m.	March
	8.00 p.m.	May		7.00 p.m.	November
June	10.00 p.m.	June	December	9.00 p.m.	December
	0.00 a.m.	July		11.00 p.m.	January
	2.00 a.m.	August		1.00 a.m.	February
	4.00 a.m.	September		3.00 a.m.	March
	6.00 a.m.	October		5.00 a.m.	April

The times mentioned in this table take into account the time zones in force in the European Union: winter time – from the last Sunday in October to the last Sunday in March; summer time – from the last Sunday in March to the last Sunday in October. Summer time is one hour ahead of winter time.

Example of how to use this table. The constellation Scorpius is visible to the south (in mid-July), at around 10 p.m., as shown on Map M7. Table 4.1 tells you that if you choose to observe the constellations in July, but, say, at 4 a.m., you would have to use the October indication (Maps M4 and M6). The constellations that can be observed on the southern horizon are Pisces Austrinus and Aquarius, not Scorpius.

Note that 9 p.m. and 10 p.m. are rounded up, but also perfectly adequate for the purposes of recognising the constellations in the sky. A more rigorous indication of these times depends on the place in which the observer is viewing, more precisely the longitude of this place. In Lisbon in Portugal, for example, the time should really be 9.37 p.m. – 10.37 in the summer. For Elvas, the easternmost city in Portugal in Portugal, the time would be more like 9.29 p.m. (10.29). But if you want to be reasonably strict, 9 p.m. can be considered 9.30, and 10 can be considered 10.30. Even so, the reader can gloss over this note in passing, as none of this will interfere with identifying the constellations.

Facing south, the reader will have east to the left and west to the right. Each rectangular map covers a width of 100°, from east to west, equivalent to "five spans in the sky". In this way, the sky visible on any given occasion does not fit on one map. The parts of the sky that are further to the left (east) and to the right (west) than those which appear in each rectangular map are covered by two of the remaining rectangular maps, which make the continuation. The connection is made by matching the names of the months. The circumpolar stars at a latitude of about 40°N can be seen on Map M1.

There is a process you can follow to make sure that the appearance of the constellations in the rectangular maps continues to concur with what can be seen in the sky. When you use a map to show part of the sky seen to the left of the south (between south and east), you should turn it to the left, as shown in Figure 4.3. If you are looking at the map that shows what can be seen to the right (between south and west), it is useful to move it round to the right, also shown in Figure 4.3. There is a simple explanation for this: we have seen that the plane of the celestial equator is inclined in relation to the horizon and reaches its maximum height to the south (see Figure 3.15 and 3.16). Consequently, each constellation takes on a differently inclined aspect in relation to the horizon, as the night goes on and the constellations move westwards (see Figures 3.7 and 3.10).

It is easy to know if we are inclining the map sufficiently. The idea is to make the map match what you see in the sky. Having identified two relatively bright stars, lean the map in such a way that an imaginary line linking the stars on the map has approximately the same angle as the one that you can see "drawn in the sky" between these same stars

In Figure 4.3, all of the "S" arrows point towards the celestial south pole, below the horizon for us in the northern hemisphere. All "N" arrows point towards the celestial north pole, although it may not appear to be the case, since we see them moving away from each other. But oranges, for example, are approximately spherical, and the lines between orange segments also move out from one of their poles only to meet again at the other. This is something we understand without difficulty and is exactly what happens in the celestial sphere. However, a shape drawn on a page has to be flat (as in Figure 4.3), which leads to a certain number of distortions.

In the bottom part of each of the rectangular maps you will find the words "visibility limit for latitude 40° N". From the central belt of the USA, one cannot see stars further south of this point, as they are always obscured by the horizon. It is worth noting that any

Figure 4.3. Positions for map orientation, in accordance with the direction of the horizon one is looking at. The letters N, S, E and W stand for north, south, east and west respectively.

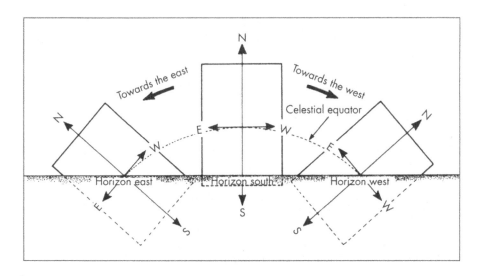

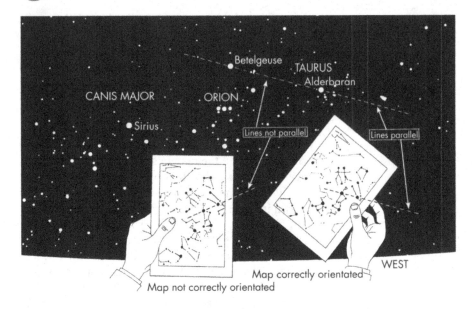

Betelgeuse

TAURUS

Alderbaran

CANIS MAJOR

ORION

Lines not parallel

Lines parallel

Sirius

WEST

Map correctly orientated

Map not correctly orientated

star at less than 5° above the horizon at any given time appears much fainter (see Section 2.8). For this reason a third-magnitude star is not usually visible to the naked eye if it is very close to the horizon, regardless of how good the place of observation may be. You therefore never have a decent view of celestial regions close to the horizon. All readers will have seen that when the Moon rises it is less bright than when it is high in the sky, which also happens, somewhat more obviously, with the Sun.

In each Identification Map, reference is made to the starting point, which will be a constellation or a bright star previously identified in maps and diagrams. In rectangular Maps M2 to M7, one could start on a different constellation that may be easy to identify on the southern horizon.

On the outside of each rectangular map, the reader can find indications **Eq** (on the left and right) and **Z** (only on the right). If you place a ruler on the lines of the two **Eq**'s, the ruler will indicate the celestial **equator**. If you place your ruler parallel to the base of the maps, passing along the line of **Z**, the ruler will show the stars that pass through the zenith at a latitude of 40° N, the average latitude of the central belt of the USA. Above the **Z** point, you have gone past the zenith. So that you don't have to twist your head too far round, turn the map the other way when you want to look at the parts of the sky north of the zenith. One of the best ways of

Figure 4.4.
Orientation of a map in relation to the sky, taking two stars as reference. In this case, the stars used are Betelgeuse (α Orionis) and Aldebaran (α Tauri). Naturally, the stars you use can be from the same constellation or different constellations. Circular Maps M1 and M8 are oriented in the same way.

contemplating the sky, without getting tired and with the best possible field of vision, is lying on the ground. With your head facing north and your feet pointing south, you can see the entire map and compare it with the sky without having to turn the page around.

The indications on using the circular maps are to be found in the text accompanying the maps. In order to make connections between maps easier, the circular maps contain some further stars outside the circle, in order to dovetail with the square maps.

It is neither necessary nor desirable to use binoculars or telescopes to learn to identify constellations and the brightest stars. Once you get to know how to identify, with confidence, some nine or ten constellations in different positions relative to the horizon and to recognise easily one or two stars in each of these constellations, then you can start to use binoculars. Having said that, it is not with binoculars that you will learn to identify further constellations. In other words, by picking up binoculars you do not completely drop naked eye observations, which are essential to learning further constellations.

After Map M7, there are three detailed maps which enable you to locate certain interesting "objects" with binoculars: an open cluster of stars, a galaxy and a nebula (see Figures 4.7, 4.8 and 4.9). Although this book is aimed at naked-eye observation I decided to include these objects for two reasons: firstly, they are detectable to the naked eye if the observer is especially alert; secondly, the objects remind us that *knowledge of the constellations is necessary* if you wish to locate these and other targets, which are accessible to observers with binoculars or a telescope. Indicating celestial objects that are not observable with the naked eye is of course outside this book's remit. The maps in Chapter 8 offer an impression of all of the sky that can be seen in each of the seasons of the year. The first four are aimed at observers in the northern hemisphere and were designed for the average latitude in central USA, 40° N.

Indication of Names on Identification Maps

The Identification Maps indicate the names of the brightest stars in order to make it easier to identify them again in future. Where necessary, I shall use these to indicate some directions in the sky that form alignments that are important for identifying other stars and some constellations. The names of stars are always in lower case (although the first letter is a capital). In some cases it is impossible to avoid other stars appearing on top of the name. In such cases, the name corresponds to the brightest star, i.e. the one represented by the largest dot. Appendix 4 offers information regarding stars whose names are indicated on the maps or in other diagrams in this book.

The names of the constellations are always written in CAPITAL LETTERS, in order to avoid any confusion with the names of stars. I have said earlier that some constellations are easier to identify than others. Some are more difficult because their stars are fainter; others because they can only be seen partially from central USA (close to the horizon); others are difficult to identify because they are in particularly highly populated areas of the celestial sphere, where it is more difficult to make out individual constellations. Appendices 5 and 6 explain various aspects related to the visibility of the constellations and how easy they are to identify or otherwise.

4.4 Ursa Major: The Best Starting Point

Due to a fortunate cosmic coincidence the seven brightest stars of Ursa Major are ranged in the sky in such a way that they can be used to locate various bright stars and to identify corresponding constellations (see Figure 4.5). These seven stars make easy alignments and can be used on any night of the year at any time. In central USA, for someone north of New York City, they never set. In other regions of the country, when Ursa Major is at its lowest, only *Alkaid* dives below the horizon. Even so, not all the stars on this map will be visible on the same occasion.

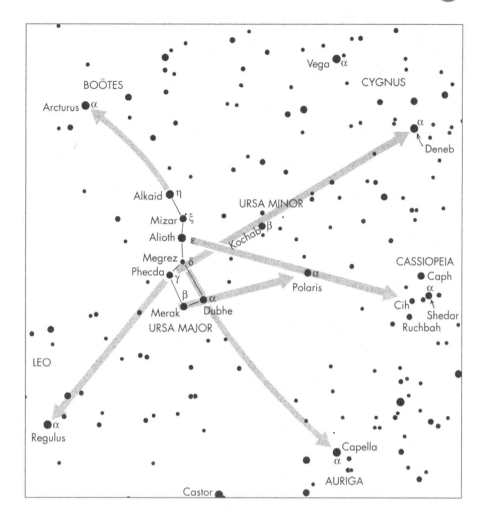

Figure 4.5.
Identification of various bright stars, using Ursa Major as a starting point. These stars, located by means of alignments and extensions shown in this diagram, will serve as a basis for the exploration of each of the following Identification Maps.

We have seen that it is possible to adjust a map according to what you can see in the sky (Figure 1.8). It is therefore possible to follow Figure 4.5 so that the appearance of Ursa Major, in the diagram, corresponds to what you see in the sky. Identification Map M1 shows you how this is done, bearing in mind the month and time of observation.

Using Figure 4.5, some of the alignments indicated will be visible and can be put into practice straight away. Even without measuring angles, you will find the stars mentioned – you cannot miss them.

If you imagine an extension of a line drawn straight from *Merak* to *Dubhe*, you will find *Polaris*, as mentioned above. Once *Polaris* has been identified, locating Ursa Major becomes relatively easy. Start by identifying *Kochab*, the second brightest star in this constellation and *Pherkad*, the third. Using Figure 4.5 or Map M1 will make this easier, and then later looking at the sky. Note that Kochab is almost as bright as *Polaris* and that Pherkad is brighter than *Megrez*, in Ursa Major. From these three stars it will be possible to imagine the positions of the remainder in the sky, and see them if you are in a very good viewing location.

Extending the arc of Ursa Major's tail, following its curve, you reach the orange coloured star *Arcturus*, the brightest in Boötes.

The alignment defined by Megrez and Phecda points towards *Regulus*, in Leo. The other way, i.e. from Phecda to Megrez takes you to *Deneb*, the brightest star in Cygnus. This alignment also passes Kochab, the second brightest star in Ursa Minor.

Extending an imaginary line straight from Megrez to Dubhe allows you to find *Capella*, the brightest star in the constellation Auriga. Capella can also be found by connecting Dubhe with *Polaris*; when you get to *Polaris*, turn 90° to the right. A further alignment, from Alioth to *Polaris*, shows the location of Cassiopeia, the two brightest stars of which are *Shedar* and *Caph*.

You can extend a line drawn from Megrez to Merak, leading to *Castor* and *Pollux*, the brightest stars in Gemini. Castor is visible towards the bottom of Figure 4.5.

If you are feeling lost, and do not know where an alignment (or line extension) finishes, Table 4.2 shows the angular distances between the stars in Figure 4.5, which the reader can use for guidance. You can measure these angles approximately yourself, using your hand at arm's length, as shown in Figures 3.19, 3.20 and 3.21.

These angular distances between stars are shown for information purposes, for the order in which they show the alignments and as such, they may become useful. There is no obligation for the reader to make use of these distances. You can start perfectly well by identifying the constellations and use this data later on. You can start using the Identification Maps on the following pages, with the alignments in Figure 4.5 and using the indicated stars and constellations as a starting point.

The reader will find repeated references, throughout the Identification Maps, to Ursa Major and its seven

Table 4.2. Angular distances between some stars close to the celestial north pole

Between stars	Angular distance
Dubhe (Ursa Major) and North Star (Ursa Minor)	28°
Alkaid (Ursa Major) and Arcturus (Boötes)	30°
Phecda (Ursa Major) and Regulus (Leo)	47°
Megrez (Ursa Major) and Deneb (Cygnus)	68°
Dubhe (Ursa Major) and Capella (Auriga)	49°
North Star (Ursa Minor) and Caph (Cassiopeia)	30°
Merak (Ursa Major) and Castor (Gemini)	43°

brightest stars. This repetition is intentional. As we shall see, these stars are the key to identifying numerous other stars and constellations. It is worth getting to know them well. They will gradually become familiar and, almost without thinking, you will end up knowing all their names.

Appendix 8 shows the angular distances between some stars visible in Maps M1 to M7.

What if it Is not Possible to See Ursa Major?

As previously mentioned, the stars of Ursa Major are almost all permanently above the horizon for observers in Central USA. In latitudes to the north of New York, the shape made by the seven most conspicuous stars in Ursa Major is circumpolar and is therefore accessible on any night of the year, provided the sky is clear.

But, when Ursa Major is very low, adjacent to the northern horizon at the beginning of autumn nights, all it takes is a slight mist or a little light pollution for Ursa Major to disappear from view. There may also be an obstacle low on the northern horizon blocking the constellation.

Figure 4.6 shows how to find Polaris without having to make use of the Pointers of Ursa Major, Merak and Dubhe. Imagine in the sky an angle made by three stars in Cassiopeia, *Caph*, *Shedar* and *Ruchbah*, with the vertex at Shedar. Another line bisecting this angle points toward Polaris. The angular distance between Cih and Polaris is around 32° (approximately one span + one fist).

Cassiopeia can also be used as a starting point, as shown on Maps M1 and M4. For example, an imaginary line in the sky, straight from Caph to Shedar points towards Almach in Andromeda. The other way, this alignment points in the direction of Alderamin, in Cepheus. A further alignment, from Cih to Ruchbah, takes you to Mirfak, in Perseus.

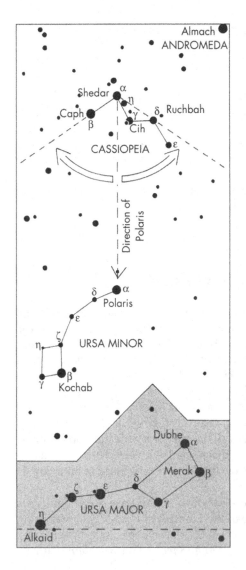

Figure 4.6. Location of Polaris when Ursa Major is not easy to see or when it is blocked by an obstacle.

4.5 Identification Maps

Identification Map 1

Circumpolar stars at latitude 40° N. Starting Point – Ursa Major

To use this map the observer should turn and face *north*, holding the book almost vertically in front of them. In order to match the map with what can be seen in the sky, turn it around so that the month in which the observation is taking place is pointing *upwards*. It is also possible to move it around according to the time of the month (beginning of the month, end, etc.). You will have the sky's appearance as it is at the beginning of the night, in each month of any year, just after nightfall (around 9 p.m. in winter time and 10 p.m. in summer time). Each night, as the hours pass, the stars rotate anticlockwise around the celestial north pole, as shown by the arrows at 15° per hour. For each hour later, the map should be turned round 15° anticlockwise, so that what you see on the map matches what you see in the sky. For each hour earlier, move the map around 15° in a clockwise direction. The dividing lines on either side of the name of the month appear at 30° intervals to make things easier for the reader.

The stars on this map, within the circle, are always visible from any place where the latitude is approximately 40° N at any time of day, at any time of year – they are circumpolar stars. The stars outside the circumference and the brightest stars are also shown, so as not to cut the sky off abruptly and so as to make it easier to connect this map to the rectangular Maps M2 to M7. The outer shape of the map is square (connection is made easier bearing in mind the names of the months). The bottom of the circular map, which at any given moment is pointing downwards, shows the stars in the vicinity of the northern horizon. If the northern horizon is obstructed, or if there is a significant amount of illumination in the area, there is a chance you will not see the stars that are just above the horizon. The upper part of the square map shows stars which, at the time of observation, are close to the zenith.

This map can be used in areas of latitude between 30° and 50°, although for other latitudes circumpolar stars do not correspond exactly to those indicated here. Where the latitude is substantially below 40° N, the diameter of the circumpolar region of the sky will be smaller and will thus encompass fewer stars. If the latitude of the viewing location is considerably higher than 40° N, then the reverse happens.

Let us look at how to find, for example, the constellation Draco, the Dragon. The stars of this constellation are not especially bright, with the exception of Eltanin (γ Draconis), which is situated at the head of the mythical beast. It is therefore a good idea to find this star first. In order to do this Polaris must be present. Start by "drawing" a straight line from Megrez (δ Ursae Majoris) to Polaris, then turn 90° to the left (in relation to the path you were following). In this new direction follow approximately the same angular distance as between Megrez and Polaris. The brightest star in this area of the sky is Eltanin. Once you have found this star, you will be able to see the quadrilateral that makes the Dragon's head. In turn, you will now see that the other, less bright stars in this constellation snake their way between the two bears, Ursa Major and Ursa Minor. In Map M6, you can find a different path to the Dragon's head, from Vega, in the constellation Lyra.

Alkaid (η Ursae Majoris), Polaris and Eltanin (γ Draconis) make a triangle of approximately equal sides, which is easy to recognise.

You will also find the names of various stars in Ursa Minor and Cassiopeia, so that they can be referred to in future alignments. For example, following the alignment defined by Cih and Ruchbah (both Cassiopeia), you can find Mirfak, the brightest star in the constellation Perseus. The alignment defined by two other stars in Cassiopeia, Caph and Shedar, leads to Alderamin, the brightest star in Cepheus. Alderamin is white and slightly fainter than Polaris. It can also be found by following the alignment from Alioth (ε Ursae Majoris) to Kochab (β Ursae Minoris). The brightest stars in Cepheus make a pentagon shape in the sky, in which Alderamin stands out and in which Errai can also be found, if you follow an alignment drawn from Dubhe (α Ursae Majoris) to Polaris, at around 11° from the latter.

If you extend Ursa Major's tail you will find Arcturus, the most evident star in Boötes. It is an extremely bright star, yellow in colour. This alignment is not marked on this map (please see also Figure 4.5).

Taking the path that has already led you from Dubhe to the North Star (Polaris) and beyond, you will eventually reach The Great Square of Pegasus, as shown in Maps M4 and M5. This alignment is detailed in Figure 6.4. The distance between Polaris and the Great Square is around 65°, i.e. a little less than three spans.

The constellations Lynx and Camelopardalis (Giraffe) pass almost unnoticed, as they only have faint stars. They can be identified with the aid of this map, but you will need a particularly dark sky and very good eyesight.

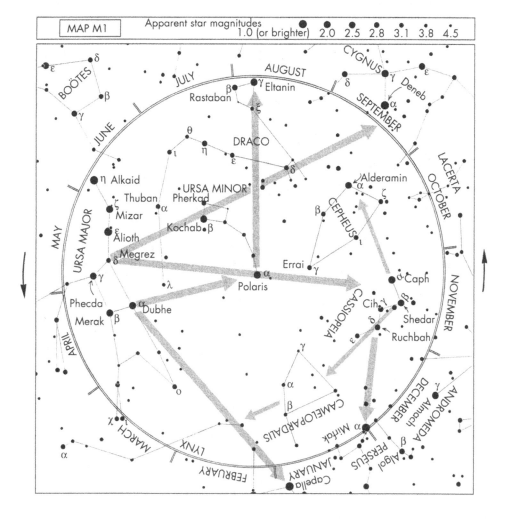

Identification Map 2

Starting Point: Ursa Major

This area of the sky is visible on spring nights. Let us use the curve of The Great Bear's tail as a guide. Following this curve, you will find Arcturus, the brightest star in Boötes, conspicuous in the region with its orange hue. The angular distance between Alkaid, at the bottom tip of the tail, and Arcturus is around 30°. After Arcturus, 30° on, continuing the curve, you will find Spica, bluish-white and the brightest star in Virgo. A little further on, you can see a small quadrilateral of stars that are faint, but relatively easy to find. These define the constellation Corvus.

The alignment drawn by Merak and Phecda in Ursa Major takes you to Regulus, the brightest star in Leo. The configuration of the brightest stars in this constellation clearly draws the shape of a lion. At the tip of the tail, you can find Denebola (β Leonis), a second-magnitude star. This is a large and imposing constellation, with Regulus practically in line with the ecliptic.

Leo faces west and has Cancer in front of it, one of the faintest constellations in the zodiac. The lion's tail points towards Virgo, where, in addition to Spica (α Virginis), there are two other conspicuous stars.

Continuing along this alignment, from Phecda to Regulus, you come to Alphard, the brightest star in Hydra (second magnitude), which is orange in colour, and is situated in a bright triangle usually known as the Spring Triangle.

Orange in colour, it was said in ancient times that this star represented the heart of this many-headed mythical beast killed by Hercules. The main head, considered immortal, was marked by a ring of the least bright stars visible on this map. An extension of 100° from the head to the tip of the tail makes Hydra the longest constellation in the sky. It is not, however, especially conspicuous. Alphard (α Hydrae) and the ring of stars representing the head are easy to see, but the remaining stars are not very bright.

The alignment defined approximately by Megrez and Merak (Ursa Major) points towards Gemini, shown on the next map, in which the two brightest stars, Pollux, blue-white, and Castor, orange-white, can clearly be seen.

In this area of the sky there are other constellations in which the only stars are very faint, making them difficult to identify. Crater, adjacent to Hydra and Corvus, and Sextans, close to Crater, come into this category. Leo Minor, between Leo and Ursa Major, is also a very faint constellation, as is Coma Berenices, which lies between Boötes and Leo. Canis Venatici is practically halfway along a line drawn from Arcturus (α Boötis) to Merak (β Ursae Majoris). The brightest star in this constellation is Cor Caroli, which can be found easily by imagining a line from Mizar to Alkaid (Ursa Major), then turning 90° left. Cor Caroli is brighter than Megrez.

Further down, almost brushing the southern horizon, are the constellations of Centaurus and Vela. These can only be seen partially the central belt of the USA, as they do not appear much above the horizon. For those in the southern hemisphere (or in the northern hemisphere, but further south than central USA), these constellations appear higher up in the sky and their stars shine brightly. In such conditions, Centaurus is very imposing.

Between 1750 and 1780, the French astronomer Nicolas Louis de Lacaille divided up the old constellation of the Ship, which had been considered much too big, into four smaller constellations: Vela (sails) and Circinus (drafting-compass), on this map, and Carina (keel) and Puppis (poop), on Maps M3 and M8.

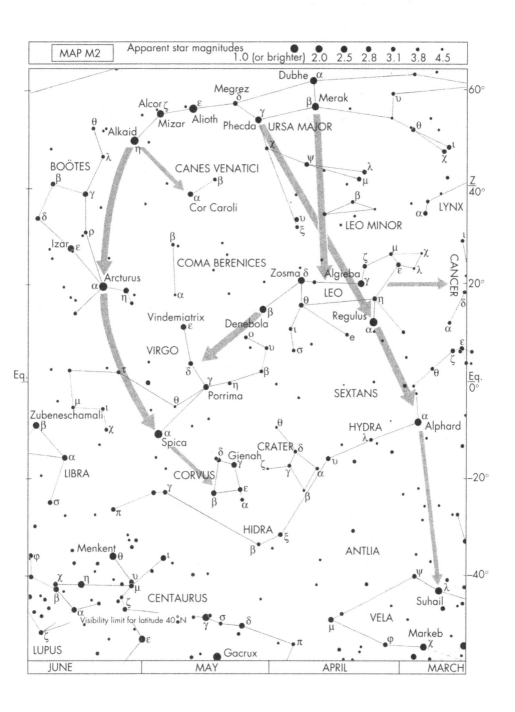

MAP M2

Apparent star magnitudes
1.0 (or brighter) 2.0 2.5 2.8 3.1 3.8 4.5

JUNE · MAY · APRIL · MARCH

Identification Map 3

Starting Points: Capella (Boötes) and Orion, the Hunter

The areas of the sky visible during the winter contain more bright stars than at any other time of year. Following an alignment from Megrez to Dubhe, in Ursa Major (Map M1), you will find Capella, the brightest star in Boötes (yellow-white), close to the top of this map.

The constellation Orion is the brightest in this part of the sky. Of the 50 brightest stars in the sky, six belong to Orion. This constellation was named after a hunter with a stick facing a bull. It is vaguely rectangular, with Betelgeuse top left, a red supergiant star 300 times greater in diameter than the Sun. In the opposite corner you will find Rigel, a blue-white star. Betelgeuse and Bellatrix (the third-brightest star in the constellation) marked out the positions representing the hunter's shoulders. One of the most obvious characteristics of this constellation is the existence of three second-magnitude stars – Alnitak, Alnilam and Mintaka (see Figure 4.9), spaced evenly apart. They are known as Orion's Belt. Below this belt is the Great Nebula of Orion, visible with binoculars in a dark location (see Figure 4.9). The celestial equator passes though this constellation, close to Mintaka (δ Orionis).

Extending the direction defined by the three stars that make Orion's Belt, you will meet Sirius, the brightest star in Canis Major and the one which stands out most in the sky. In this constellation, Sirius is 8.6 light years away and Wezen (δ Canis Majoris), which seems to be alongside Sirius, is in fact 2000 light years away, showing just how appearances can be deceptive. After Sirius, turn right 90° and you will meet Procyon, the principal star in Canis Minor. Betelgeuse, Sirius and Procyon make an almost equilateral triangle in the sky. In turn, Procyon (α Canis Minoris), Regulus (α Leonis) and Alphard (α Hydrae) make a right-angled triangle, with the right angle at Alphard. Moving along beyond Orion's Belt in the other direction takes you to Aldebaran, the brightest star in Taurus. It is orange in colour and used to represent one of the animal's eyes. Taurus the bull's horns are also clearly visible. At the tip of one of the horns is Elnath, the second brightest star in the constellation (see Appendix 1). Carrying further on, and arcing slightly right, still in Taurus, is the open cluster of stars known as the Pleiades, which is visible to the naked eye, but which is much more interesting through binoculars (see Figure 4.7).

Approximately halfway along a line from Capella (α Aurigae) to Regulus (α Leonis), you can find the two brightest stars in Gemini, Castor and Pollux. It is also possible to locate Castor and Pollux if you go from Sirius to Procyon, bending slightly to the right, i.e. the side where Capella is situated. Here is a tip for avoiding confusion between these two stars in Gemini: Castor is close to Capella (both begin with C) and Pollux is on the same side as Procyon (letter P). The alignment defined by Castor and Pollux points to Alphard (α Hydrae). Between Gemini and Ursa Major lies Lynx, a very faint constellation. The seven brightest stars Rigel (β Orionis), Sirius (α Canis Majoris), Procyon (α Canis Minoris), Pollux and Castor (α and β Geminorum), Capella (α Aurigae) and Aldebaran (α Tauri) make a rough circle in the sky, sometimes known as The Winter Circle, as they are most visible in the winter at the beginning of the night. Notice that Capella can also be found by following a line drawn from Regulus to Pollux. At Orion's feet is the constellation Lepus, whose brightest star is Arneb (α Leporis). To the south of Lepus, in the direction out from the back legs of Canis Major, is the faint constellation of Columba, the Dove. Further on, you arrive at the constellation of Caelum, the Chisel, which only has very faint stars. Behind Canis Major is Puppis (poop of the Ship), and then Circinus, the drafting compass and Vela, the sails. Puppis and Vela are scarcely above the horizon and only the upper part of Puppis can be seen from central USA. Circinus only has faint stars, making it difficult to recognise.

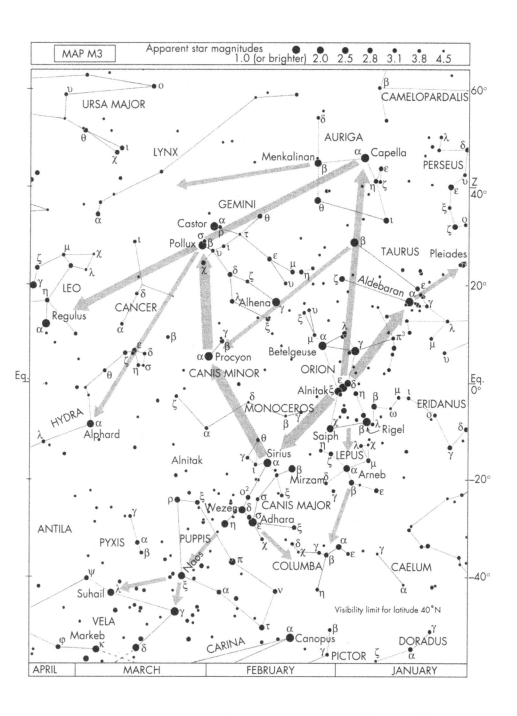

Identification Map 4

Starting point: Cassiopeia

Autumn is the time of year in which fewest stars can be seen. But by way of compensation the nights are relatively calm and long, offering excellent conditions for observing the sky. Cassiopeia has already been identified in previous diagrams and is now our starting point for recognition of this part of the sky, in Map M4. Extending the line from Cih to Shedar and beyond, in Cassiopeia, you will find The Great Square of Pegasus, an unmistakable figure, marked by four stars of magnitude 2 to 3, Sirrah, Scheat, Markab and Algenib, of which the first belongs to Andromeda and the other three to Pegasus, the mythical winged horse of Greek mythology. The Great Square is not a constellation, but a geometric figure that can be used as a reference point. It "measures" about 14° × 17° and resembles more a trapezium than a square. Once you have located the Great Square you can quickly find the constellations of Pegasus and Andromeda and, using its sides and diagonals, you can go exploring. The diagonal of this square drawn from Algenib to Scheat points to Deneb (α Cygni), a star already defined in Figure 4.5 and shown in Maps M1 and M5. The Great Square can also be found from the Pointers of Ursa Major and Polaris, as mentioned in the text accompanying Map M1. The direction from Markab to Scheat leads you to Polaris.

Sirrah is the brightest star in Andromeda. In this constellation the Galaxy of Andromeda is also visible, 2.9 million light years away, which at this time of year is well above the horizon and in the best observation conditions (see Figure 4.8). The other two brightest stars in Andromeda are Mirach and Almach.

The constellation Aries, which looks nothing like a ram, can be identified, provided you can spot its brightest star, Hamal, which makes an isosceles triangle with two stars of the Great Square of Pegasus, Sirrah and Algenib. You can also find Hamal by drawing a straight line from Sirrah to Aldebaran (Taurus); part way between these stars is Hamal. The second brightest star in Aries, close to Hamal, is Sheratan.

Between Almach (γ Andromedae) and the constellation Aries is Triangulum, a small constellation with faint, but unmistakable stars that are reasonably easy to identify, provided that viewing conditions are satisfactory.

Extending the direction defined by Sirrah and Algenib, the Great Square brings you to Diphda, a second-magnitude star situated at the tip of the tail of Cetus, the Whale. This extension takes you across the constellation Pisces, whose stars are faint and which represented two fish connected at the tail, at the spot marked by Alrischa (α Piscium), another faint star. The spring begins (northern hemisphere) when the Sun passes in the direction of Pisces. The star Mira is also in Cetus and its magnitude varies between 3 and 9 in a period of 331 days. It is only visible to the naked eye when it is at its brightest.

Between Cassiopeia and Auriga is Perseus, the Greek mythological hero who managed to rescue the Princess Andromeda from the clutches of the whale. From Cassiopeia, you can see that the line from Cih and Ruchbah leads to Mirfak, the brightest star in Perseus, yellow-white in colour. In Perseus you will find Algol (β Persei), a notable eclipsing variable, whose brightness varies between 2.2 and 3.5 magnitudes in a period of 2.87 days. The extension defined by the three brightest stars in Andromeda, namely Sirrah, Mirach and Almach, points almost exactly towards Mirfak, which also leads you to Perseus.

Eridanus, the river of the tears in Greek mythology, "rises" adjacent to Rigel, Orion's right foot. This constellation is very large and only has one bright star, Achernar, which cannot be seen from Central USA. Next to Eridanus are the constellations of Caelum, the chisel and Fornax, the furnace, which contain only faint stars visible only under excellent viewing conditions. Further south are the constellations Horologium, the pendulum clock and Dorado, the goldfish. These are both very faint and the latter is completely inaccessible from central USA. To the south of Cetus is the constellation Sculptor, whose stars are again very faint and which goes almost unnoticed. To the south of the Sculptor and almost inaccessible from cenntral USA is Phoenix, which only has one bright star and which only grazes the horizon, as it is situated way to the south of the celestial equator.

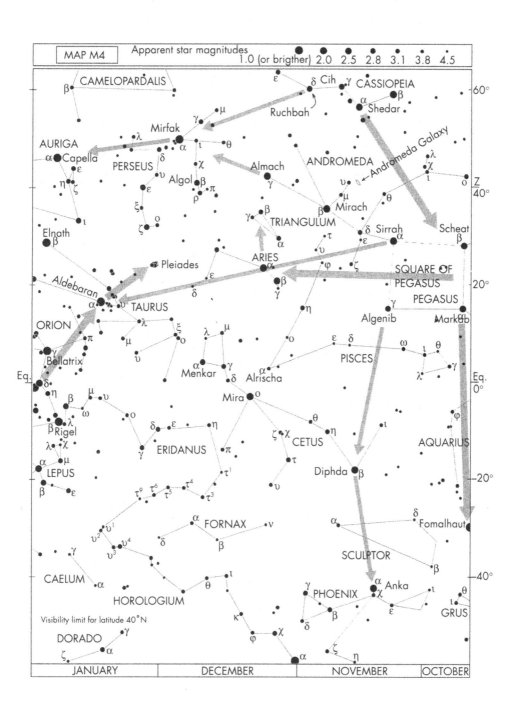

MAP M4

Apparent star magnitudes
1.0 (or brigther) 2.0 2.5 2.8 3.1 3.8 4.5

CAMELOPARDALIS

β

60°

ε   δ Cih   γ CASSIOPEIA
α   β
Ruchbah   Shedar

γ   μ

Mirfak

λ   Andromeda Galaxy

AURIGA   λ   ι   θ   ANDROMEDA   λ   χ
α Capella   δ   α   ι   ο   Z
ε   PERSEUS   χ   Almach   υ ο   40°
η ζ   Algol β   γ   μ   θ
ξ   π   β Mirach
ρ   δ Sirrah   Scheat
Elnath   ζ   ο   γ β   β   α   β
β   TRIANGULUM   τ
ARIES   υ   SQUARE OF
Pleiades   α   φ ζ   PEGASUS   20°
Aldebaran   ε   PEGASUS
δ   β   γ   Algenib   Markab
α   υ   η
ORION   TAURUS   λ   ξ   ο   Algenib   Markab
λ   μ   ο   PISCES   ω   ι   θ
Bellatrix   υ   ε δ   λ   γ
Eq.   α   Menkar   γ   α Alrischa   Eq.
δ   η   β   μ υ   ο   Mira ο   φ   0°
ω   δ ε   η   θ   AQUARIUS
β Rigel   δ ε   η   ζ χ   η   ι
λ χ   γ   ERIDANUS   π CETUS
α   μ   π   τ
LEPUS   Diphda β
β   ε   τ⁹ τ⁶ τ⁵ τ⁴   Fomalhaut
τ³   α   δ
υ²υ¹   α FORNAX ν   20°
γ   υ⁴   δ   α
υ³   β   δ Fomalhaut
CAELUM   α   ι   SCULPTOR   40°
β
HOROLOGIUM   γ α Anka
Visibility limit for latitude 40°N   PHOENIX χ   ι   θ
κ   ε   GRUS
DORADO   γ   φ   χ   δ
ζ   α   α   ζ   η

JANUARY   DECEMBER   NOVEMBER   OCTOBER

Identification Map 5

Starting point: the Great Square of Pegasus

The Great Square of Pegasus was identified on the previous map and is your new starting point. For this purpose, let us remember which stars make this square in the sky: Sirrah (α Andromedae), Algenib (γ Pegasi), Markab (α Pegasi) and Scheat (β Pegasi). Note that these four stars are not particularly bright, from magnitude 2.1 (Sirrah) to 2.9 (Algenib), but which are easy to see, as they are situated in an area of the sky which is not rich in prominent stars. The Square is thus easy to find and is characteristic of the autumn sky in the northern hemisphere. Sirrah, also known as Alpheratz, belongs to the neighbouring constellation of Andromeda, as mentioned in the previous map.

Coincidentally, two of the sides of The Great Square of Pegasus point almost perfectly north–south and the other two east–west. The side defined by Markab and Scheat has a north–south direction, which eventually leads to Polaris if you extend north way beyond the confines of this map. It is 60° from Scheat to Polaris. Extending one of the diagonals from Algenib to Scheat points almost in the direction of Deneb, the brightest star in Cygnus. From Scheat to Deneb, the reader's eyes will move round 33° in the sky, or one span and a half. Once you have found Deneb, you will see that the brightest stars in Cygnus draw the shape of a cross, sometimes called the Northern Cross. If you extend the small arm of the cross you will get to Enif (ε Pegasi).

Back to the Great Square and following an alignment from Markab to Enif leads to Aquila, the Eagle, where the star Altair stands out. If you extend the line which runs from Sirrah (α Andromedae) to Algenib (γ Pegasi) you will find the star Diphda, in the constellation Cetus, an extremely large constellation, extending as far as Taurus, as shown on map M4. Further south is Anka, the brightest star in Phoenix. For an observer in the central belt of the USA, this constellation does not appear much above the southern horizon.

The diagonal of the Great Square of Pegasus from Algenib to Scheat, in the opposite direction from the one taking you to Cygnus, points towards Mira, in Cetus. Notice that Mira makes a large isosceles triangle with Diphda (β Ceti) and Algenibe (γ Pegasi). However, Mira's magnitude varies between 3 and 9 in an average period of 331 days, as it is a long-term variable star. It is only visible to the naked eye at its brightest, for about six or seven weeks a year. You should therefore not be surprised if you cannot see Mira on your celestial explorations. It is a red supergiant, whose diameter is 300–400 times greater than that of the Sun. The map shows Mira at its brightest.

Extending one of the sides of the Square from Scheat to Markab, go past Aquarius, whose stars are very faint, and you arrive at Fomalhaut, the brightest star in Pisces Austrinus, the Southern Fish. Fomalhaut is approximately the same brightness as Deneb. You can see that Diphda (β Ceti), Anka (α Phoenicis) and Fomalhaut (α Piscis Austrini) make an almost equilateral triangle. The constellation Sculptor is almost entirely inside this triangle, but is very faint, its brightest star being only magnitude 4.4.

To the south of Pisces Austrinus is the constellation of Grus, the crane, only partially visible from central USA. Its brightest star, Alnair, appears only just above the horizon, in a southerly direction. A line drawn straight from Scheat (β Pegasi) to Altair (α Aquilae), passes by the small constellation of Delphinus, the Dolphin, whose stars are faint.

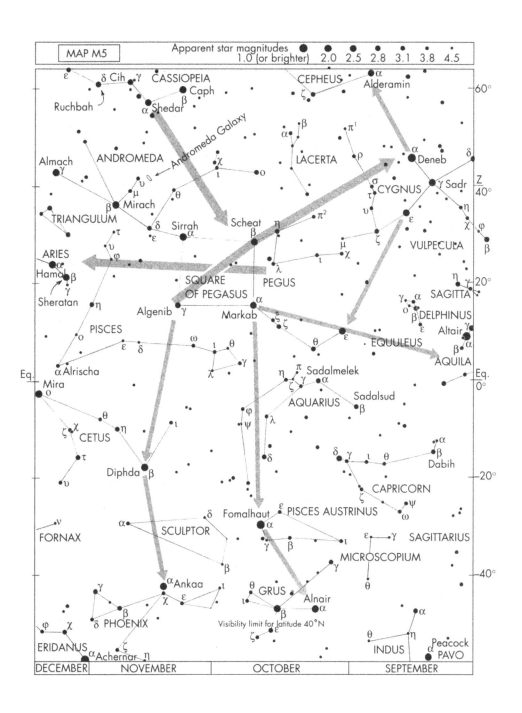

MAP M5

Apparent star magnitudes
1.0 (or brighter) 2.0 2.5 2.8 3.1 3.8 4.5

ε δ Cih γ CASSIOPEIA   CEPHEUS α
                                          Alderamin —60°
                Caph   ζ
Ruchbah   β
          α Shedar

                    Andromeda Galaxy   π¹   α
ANDROMEDA   χ   Deneb δ
Almach   ι   LACERTA ρ   Z
γ   ο   σ γ Sadr   40°
        υ ο   α   τ CYGNUS   η
      μ   θ   χ φ
  β Mirach   υ   ε   β
TRIANGULUM   δ Sirrah   Scheat η   π²   VULPECULA
      τ   ε   α   β   μ κ   ζ
ARIES υ φ   ι χ   η γ —20°
Hamal β   λ   γ α SAGITTA
  γ   SQUARE   PEGUS   ο
Sheratan η   OF PEGASUS α   β DELPHINUS
        Algenib γ Markab   ε   Altair γ
PISCES   ζ ζ   EQUULEUS   β α
  ο   ω ι θ   θ ε   AQUILA
Eq. α Alrischa   χ   γ   Eq.
Mira   η π Sadalmelek   0°
  ο   θ   ζ γ α
ζ χ   η   φ AQUARIUS Sadalsud
  CETUS   ψ λ   β
    τ   δ γ ι θ   α
  υ   β
        Diphda β   Dabih
                                        ζ
                    δ Fomalhaut ε PISCES AUSTRINUS ψ —20°
        α   SCULPTOR   α   ω
FORNAX   γ β ι   ε γ SAGITTARIUS
                          β   MICROSCOPIUM
                                              γ   —40°
              γ   α Ankaa ι θ   θ
          β   ε   GRUS ι
  φ χ   δ PHOENIX χ Visibility limit for latitude 40°N Alnair
ERIDANUS   ζ ε   θ η Peacock
  α Achernar η   INDUS α PAVO

DECEMBER   NOVEMBER   OCTOBER   SEPTEMBER

Identification Map 6

Starting point: Summer Triangle

On this map only two stars of The Great Square of Pegasus are shown – Scheat and Markab. Extending one of the diagonals of the Square, from Algenib to Scheat, takes you to Deneb, the brightest star in Cygnus, as explained on the previous map. This constellation represented a swan with its wings open, the tip of its tail at Deneb and the beak at the star Albireo. From Deneb you will see in this region of the celestial sphere two other very bright stars – Vega (α Lyrae) and Altair (α Aquilae). Deneb, Vega and Altair are very conspicuous in this area, making an (almost) isosceles triangle, which is highly prominent and is referred to as the Summer Triangle because during the summer (in the northern hemisphere), at the beginning of the night, it is very high in the sky. Of these three stars Vega is the brightest and Deneb the faintest. The smallest side of the triangle is from Vega to Deneb. Altair can also be located from Pegasus, by drawing a line from Markab to Enif. The Summer Triangle is a useful geometric figure for recognition of the sky, but is not a constellation. Each of the three stars is in a different constellation.

Eltanin, the brightest star in Draco, is almost in alignment with Altair and Vega – 26° from Vega to Eltanin. A line drawn straight from Altair to Deneb and beyond takes you to Alderamin, the brightest star in Cepheus. Extending a line from Vega (α Lyrae) to Altair (α Aquilae), you will eventually reach Capricorn, which does not contain any bright star. In the ancient world, this constellation represented a goat with a fish's tail. It is easier to spot that the brightest stars in this constellation draw a triangle in the sky with one of the vertices facing south. The Capricorn's head points west, looking at the constellation of Sagittarius, a mythical half-man, half-horse holding a bow and getting ready to release an arrow. Unlike Capricorn, Sagittarius is highly visible and its brightest stars make a shape in the sky reminiscent of a teapot, with the handle facing Capricorn. The constellation of Sagittarius is impressive, when seen from a good viewing location, but is only really spectacular when observed at more southern latitudes, from where it can be seen high above the horizon. The nucleus of our Galaxy is in the direction of this constellation. When the Sun passes in the direction of Sagittarius, the winter season begins (in the northern hemisphere). On the following map, I will indicate another means of locating Sagittarius.

Between Capricorn and Pisces is the constellation of Aquarius, where there are only faint stars and in which the Babylonians saw the figure of a man pouring water from a jug, or amphora. The water poured by Aquarius falls to the south, to the mouth of the Pisces Austrinus, or Southern Fish, at the star Fomalhaut. In ancient times, the beginning of winter in the northern hemisphere, was marked by the Sun passing into Capricorn, which was the beginning of the rainy season. This is why the names of the next constellations the Sun goes to are connected with water, i.e. Aquarius and Pisces. Adjacent to Pisces is Cetus, the Whale, and to the south of Aquarius is Pisces Austrinus, the Southern Fish, both also associated with water, although they do not come into contact with the ecliptic. Between Aquila, the Eagle, and Cygnus, the Swan, is the constellation Sagitta, the arrow, which is small and only has faint stars. It is very reminiscent of an arrow; according to Greek mythology, this arrow was shot at the eagle and the swan but missed the target. Vulpecula, the Fox, which is fainter still, is located between Sagitta and Cygnus. On this map, Vulpecula's stars are not connected by lines. Extending a line from Scheat (β Pegasi) to Altair (α Aquilae) leads you to the constellations of Ophiuchus and Serpens. A further line, from Deneb to Vega, leads eventually to the constellation Hercules, enormous but very faint (see Map M6). The same alignment in the opposite direction – from Vega to Deneb – takes you to Lacerta, a constellation which only contains faint stars.

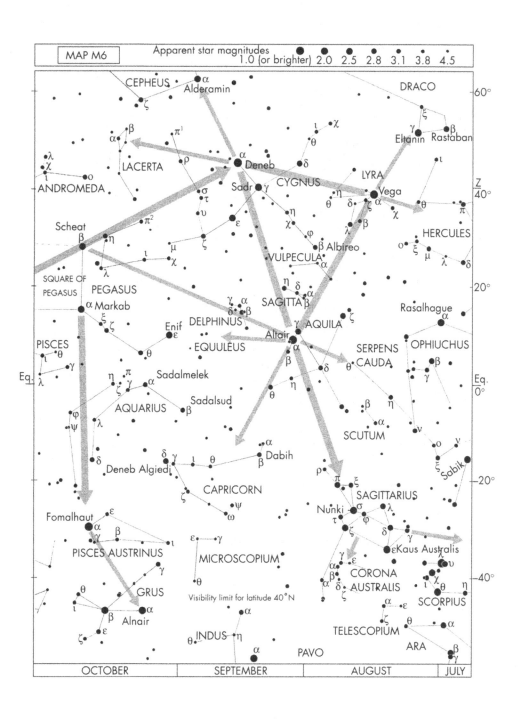

MAP M6

Apparent star magnitudes
1.0 (or brighter) 2.0 2.5 2.8 3.1 3.8 4.5

CEPHEUS
α Alderamin
ζ
DRACO
60°

β
α
π¹
ξ
γ β
Eltanin Rastaban
λ
χ
ι
o
LACERTA
ρ
α Deneb
δ
LYRA
ι
Z
40°
ANDROMEDA
σ
Sadr γ
CYGNUS
Vega
τ
η δ α
χ
π
θ
υ
η
φ
λ β
HERCULES
Scheat
β η
π²
ε
χ
β Albireo
o ξ μ λ δ
μ ζ
VULPECULA
α
ι
χ
λ
SQUARE OF
PEGASUS
PEGASUS
η δ α
SAGITTA β
20°
γ α
Rasalhague
α Markab
ξ ζ
δ β
SAGITTA β
α
Enif
DELPHINUS
γ AQUILA
ζ
OPHIUCHUS
PISCES
ε
Altair
SERPENS
ι θ
EQUULEUS
α
CAUDA
β
Eq.
0°
γ
θ
β
δ
θ
β γ
λ
η π
α Sadalmelek
δ
Sadalsud
β η
AQUARIUS
β
α
SCUTUM
o ν
φ
θ
α
ξ Sabik
ψ
δ γ ι θ
β Dabih
ν
δ
Deneb Algiedi
ρ
π ξ
20°
CAPRICORN
SAGITTARIUS
ζ
ψ
Nunki σ λ
Fomalhaut
ε
ω
τ φ
α
β
ε γ
ζ δ γ
γ
PISCES AUSTRINUS
ι
MICROSCOPIUM
γ ε
εKaus Australis
γ
υ
θ
θ
CORONA
χ
40°
GRUS
Visibility limit for latitude 40°N
β α δ AUSTRALIS
ι θ η
β Alnair
ζ
SCORPIUS
ζ ε
α ε
α
ζ θ α
θ INDUS η
TELESCOPIUM
ARA β
α
PAVO
γ

OCTOBER | SEPTEMBER | AUGUST | JULY

Identification Map 7

Starting points: Cygnus and the Summer Triangle

On the previous map we saw how to locate Cygnus and the Summer Triangle.

The direction via Deneb (α Cygni) and Vega (α Lyrae) leads to Hercules, large but without any particularly bright stars, where the shape of a trapezium is drawn by four third-magnitude stars. In this constellation is the red star Rasalgethi (α Herculis), whose brightness varies between third and fourth magnitudes. It is one of the largest known stars, a supergiant red star with a diameter 600 times greater than that of the Sun. At its brightest it is the brightest star in its constellation. Between Hercules and Boötes, and almost on a line drawn straight between Vega (α Lyrae) and Arcturus (α Boötis), you will find the constellation Corona Borealis, the Northern Crown, unmistakable with its semicircular shape. Gemma is the brightest star in Corona Borealis and is only slightly less bright than Polaris.

The line from Deneb (α Cygni) to Altair (α Aquilae) leads to the constellation Sagittarius, whose brightest stars look more like a teapot than the mythical figure after which it was named (see Appendix 1). The two brightest stars in Sagittarius are Kaus Australis and Nunki.

Extending a line from Deneb, passing halfway between Vega and Altair, helps you to find Antares, the principal star in Scorpius, red and dominant in this region of the sky. The angular distance between Deneb and Antares is around 95°. The brightest stars in Scorpius do look remarkably like a Scorpion. For ancient peoples, Antares was in the Scorpion's heart. Scorpius is easy to find when you look directly above the horizon, in a vaguely southerly direction, at the beginning of the night in July. To the east, i.e. left, of Scorpius is Sagittarius, which we have already located via a different route, from the Summer Triangle. To the south of, i.e. below, Sagittarius and to the left of Scorpius is Corona Australis, the Southern Crown, a constellation with few bright stars. To the west of Scorpius is Libra, whose brightest star is *Zubeneschamali* (β Librae) and in which only three or four stars can be seen. The configuration of the stars barely, if at all, resembles the scales that gave their name to the constellation, with the two scales originally said to be facing Scorpius. Moving on from Scorpius, via Libra, we arrive at Virgo, whose main star is Spica. To the south of Scorpius are the constellations of Lupus and Centaurus. Viewed from central USA these constellations barely reach above the horizon.

The Summer Triangle helps you to find Rasalhague, the brightest star in the constellation Ophiuchus, the Serpent Holder, a man with a snake wrapped around his belt. Rasalhague, Vega and Altair make an almost equilateral triangle in the sky. Rasalhague is less bright than Deneb. The constellation of Serpens is the only one to be divided into two, the head and the front half to the west of Ophiuchus, whose brightest star, Unukalhai, much less bright than Polaris, is situated at the snake's neck. The snake's tail, "interrupted" by the Serpent Holder, is less bright than the front half. Notice that Rasalhague is almost on a line drawn between Altair (α Aquilae) and Arcturus (α Boötis), but closer to Altair. Arcturus (α Boötis), Gemma (α Coronae Borealis) and Unukalhai make an almost equilateral triangle in the sky.

This map shows more constellations, albeit with very faint stars, that the reader can find from previously mentioned constellations, or via other routes.

The map continues west (right) onto Map M2. The stars situated to the south of those found on Maps M2 to M7 are shown in Figure 5.2 and on Map M8, in Chapter 5.

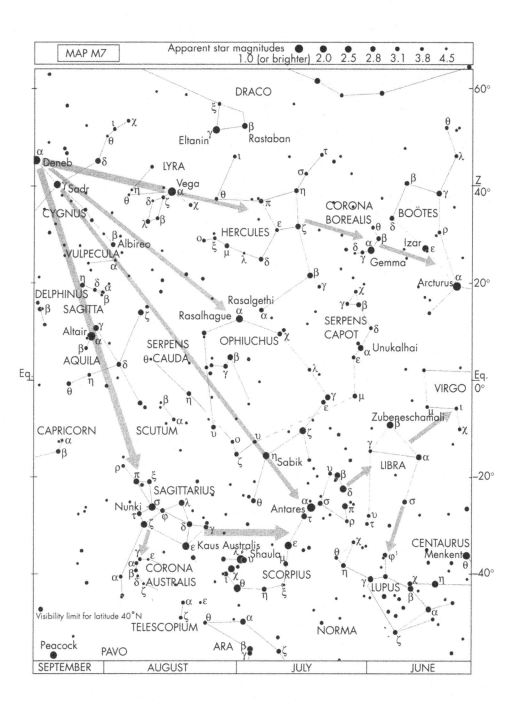

MAP M7

Apparent star magnitudes
1.0 (or brighter) 2.0 2.5 2.8 3.1 3.8 4.5

DRACO

ξ

β
γ Eltanin   Rastaban
ι

ι

χ
θ

LYRA

Vega
α

θ

σ

η

τ

CORONA
BOREALIS   BOÖTES

β

γ

Z
40°

α
Deneb   δ

60°

γ Sadr

CYGNUS

θ   δ   ζ   χ

λ   β

β
Albireo

VULPECULA
α

η   δ   α
DELPHINUS   β
β   SAGITTA

Altair   γ
β   α
AQUILA   δ

θ   η

π

HERCULES

o   ξ   μ
λ   δ

ε
ζ

Rasalgethi
Rasalhague   α
α
OPHIUCHUS   χ

SERPENS
θ•CAUDA

β

γ

θ   δ
α   β
Izar   ε

δ   β
γ   Gemma

ρ

Arcturus   α

β
γ   χ
γ•   β
SERPENS
CAPOT   δ
α Unukalhai
ε

20°

CAPRICORN
α
β

ζ

SCUTUM

α

β   η

ν
ζ

υ
o
ζ

η Sabik

λ

γ   μ
ε

Zubeneschamali
β   μ   ι
χ

γ
β   α
LIBRA

VIRGO   0°   Eq.

Eq.

20°

ρ
π   ξ

Nunki
τ   φ
ζ   δ

γ
α   ε

SAGITTARIUS
σ   λ

γ
ε Kaus Australis

CORONA
α   AUSTRALIS
β   δ   ζ

ν   β
δ
σ   π   ρ   τ
α   σ
Antares   τ

Shaula
ι   χ   υ   μ
λ
χ   θ
η   ξ

ε

θ
η

χ
φ¹

γ   LUPUS

χ

CENTAURUS
Menkent
θ

β

α

40°

Visibility limit for latitude 40°N

Peacock
PAVO

ζ   θ

TELESCOPIUM

α

ARA   β
γ

α   ε

NORMA

ζ

ζ

SEPTEMBER | AUGUST | JULY | JUNE

Identifying the Pleiades

(Location shown in Map M3)

This is the easiest cluster of stars to observe. It is sufficiently bright not to go unnoticed, even to naked-eye observers, and as such deserves special mention. Anyone with good eyesight should be able to see six or seven stars, unaided. For this reason, the Pleiades are commonly known as the Seven Sisters. The smaller diagram (inset of Figure 4.7) shows how the Pleiades look to the naked eye. With good binoculars, it is possible to see 50 stars, as in the bigger map, and a telescope with sufficient aperture would reveal more than 150. The cluster lies within the constellation Taurus, some 415 light years away from us.

Figure 4.7 identifies the main stars of the Pleiades, which in ancient times, represented Atlas, his wife Pleione, and their seven daughters: Alcyone, Asterope,

Figure 4.7.

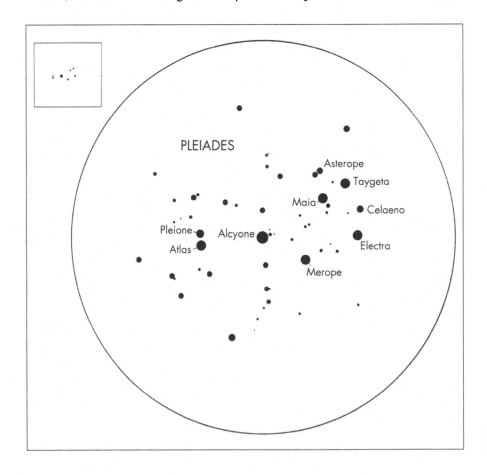

Electra, Maia, Merope, Taygeta and Celaeno. The brightest star in the cluster is Alcyone, at 2.9 magnitude. Pleione, Celaeno and Asterope are almost at the limits of what we can see unaided, with magnitudes of 5.2, 5.4 and 5.9 respectively. They can serve as a test of observing conditions and of the observer's eyesight. In the case of the Pleiades, the proximity between the stars is real and not an optical illusion and we can say that all of the stars in the cluster are practically the same distance away from Earth. The stars are bluish-white and were born together some 80 million years ago – in astronomical terms they are in fact very young. They form a system that will gradually spread out over the course of the millennia, due to the inexorable separation of the component parts. Many of the stars in the cluster are brighter than the Sun. Alcyone is 1100 times more luminous than our star and even Pleione, the least bright of the nine shown on the map, has the brightness of seventy suns. There are many clusters of this kind in our Galaxy, some of which can be seen with binoculars.

Scale: the angular distance between Atlas and Electra is 1°.

How to find the Andromeda Galaxy

See identification Map M4 for tips on locating the constellation Andromeda

Figure 4.8 shows in more detail the constellation of Andromeda and the location of the Andromeda Galaxy. At 2.9 million light years from us, this galaxy is the most distant celestial object that can be seen with the naked eye. In a moderately dark location it is visible as a whitish blot. With binoculars, its nucleus is visible, even in a city, so long as the illumination is not excessive. First find Sirrah (α Andromedae), as shown in Map M4, and then take your eye round to Mirach (β Andromedae). Then the arrows shown in this diagram indicate the route to take, either unaided or with binoculars, to find the Andromeda Galaxy. Seen through binoculars, in a good viewing location, this galaxy is spectacular. As well as the nucleus, it is possible to see a slender, luminous, oval splash around it.

What you can see is what it was 2,900,000 years ago, i.e. the light that hits our eyes came out of the stars in

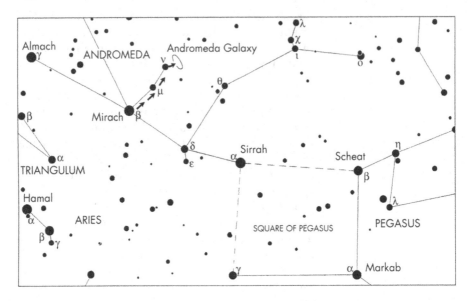

Almach
γ
ANDROMEDA
Andromeda Galaxy
ν
μ
θ
β
Mirach
β
δ
Sirrah
α
ε
TRIANGULUM
α
Scheat
β
Hamal
λ
ARIES
α
SQUARE OF PEGASUS
PEGASUS
β
γ
γ
α Markab
λ
χ
ι
ο
η

Figure 4.8.

this galaxy before the human species as we know it had walked on Earth. Many millions of galaxies are known about and some of these can be seen with a small telescope. But this one is much too bright to go unnoticed by an alert observer.

It is thought that our Galaxy, the Milky Way, described in Chapter 6, is similar to Andromeda, although the latter is slightly bigger.

How to find the Nebula of Orion

See Identification Map M3

Figure 4.9 shows in more detail the constellation of Orion and the location of the Great Orion Nebula, which is 1600 light years from us. The light that hits us left there 1600 years ago and what we see is how it was back then. It is just about detectable to the naked eye, in a dark location. It is easy to observe with binoculars, as a small milky splash, even in a moderately illuminated place. If the location is very dark and once your eyes have adapted to this darkness, the Orion Nebula is magnificent.

New stars form in this nebula, born from interstellar dust and gas, making enormous clouds illuminated by newly formed stars. This is one of several known places where stars are formed.

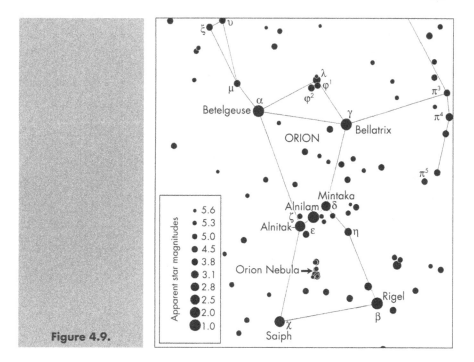

Figure 4.9.

In the diagram you can also see the stars Altinak, Alnilam and Mintaka, which form Orion's Belt and which are commonly known as the Three Marias or the Three Wise Men. The angular distance between Altinak and Alnilam is around 1.4°, approximately equal to the distance between Alnilam and Mintaka. The celestial equator passes very close to Mintaka (Figure 3.27).

The scale of apparent magnitudes, on the left, gives an idea of the brightness of each of the stars shown in this map. The magnitude of the least bright stars here is 5.6. This is why some of these stars are not indicated in Map M3.

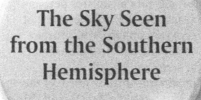

CHAPTER FIVE

The Sky Seen from the Southern Hemisphere

5.1 Other Viewing Locations

In order to help the reader to understand better the appearance of the sky seen from the southern hemisphere, and the difference from place to place on the Earth's surface, I shall start by referring briefly to the situation for observers at the equator and at each of the Earth's poles.

Observers at the Equator

People living on the equator (latitude 0°) are the only inhabitants of the Earth who have the chance to see all regions of the celestial sphere. All constellations are accessible, but not at the same time.

For these observers, the celestial poles are at the horizon: the celestial north pole is close to the horizon in a northerly direction and Polaris is adjacent to the cardinal north point of the horizon. The celestial south pole is adjacent to the horizon in a southerly direction. The celestial equator passes via the east and west points of the horizon and the zenith and the nadir. Anyone standing on the equator will be able to see all the stars in Map M8, but only half at any one time. The southern horizon is marked by one of the diameters of Map M8 and varies according to the time and date of observation. On the north side, the horizon is marked by one of the diameters of Map M1, also dependent on when the observation is taking place.

Seen from the equator, all stars make arcs perpendicular to the horizon and parallel to the celestial equator, as happens everywhere on the Earth's surface. All stars rise and set, so for the equatorial observer, there are no circumpolar stars.

The Sun reaches its maximum altitude north of the zenith, approximately between 21 March and 23 September, and to the south of the zenith, approximately between 23 September and 21 March. On 21 March and 23 September every year the Sun passes over the zenith.

Observers at the North and South Poles

North Pole

As mentioned above, for an observer standing at the north pole (latitude 90° N), the celestial north pole is located at the zenith. The celestial south pole will be at the nadir and the celestial equator at the horizon. The stars make (apparent) anticlockwise movement – due to the Earth's rotation – around the celestial north pole, following arcs parallel to the celestial equator and to the horizon. All stars visible at a given moment in time are circumpolar. Stars do not rise or set, apart from the Sun, of course. The map in Figure 6.4 shows the stars that are permanently visible to north pole observers. Using this map the observer can later identify constellations by means of the Identification Maps. For the north pole observer, any direction in which he looks from the horizon is going to be south.

The Sun is always above the horizon between 21 March and 23 September approximately (polar day), and below for the rest of the year (polar night).

South Pole

As mentioned above, an observer at the south pole (latitude 90° S) will always have the celestial south pole at the zenith. The celestial north is at the nadir and the celestial equator coincides with the horizon. The stars appear to move clockwise – due to the Earth's rotation – around the celestial south pole making arcs parallel to the celestial equator and the horizon. All stars visible at a given moment in time are circumpolar – they neither rise nor set, apart from the Sun. Using this diagram, the polar observer can later identify constellations by means of the Identification Maps. Note that for the observer at the south pole, any direction from the horizon in which he looks is always north. Figure 6.5 shows the stars permanently visible to south pole observers.

The Sun is permanently above the horizon between 23 September and 21 March, approximately (polar day) and below the rest of the year.

5.2 What Does the Sky Look Like in the Southern Hemisphere?

Seen from the southern hemisphere, the celestial sphere appears to wheel from east to west. Polaris is invisible, as it is below the horizon. In addition to the southern hemisphere observer seeing different stars and constellations from those we see in the north, as you are about to read, there are other differences which may be surprising to the reader in the northern hemisphere.

Let us see in detail how the sky is seen by the observer in the southern hemisphere, but not including the observer at the south pole. Observers at, say, the Cape of Good Hope (approximate latitude 34° S) will have the *celestial south pole* at around 34° above the *southern* horizon (see Figure 5.1). All stars less than 34° away from the celestial south pole are circumpolar. All stars up to 90° – 34° = 56° to the north of the celestial equator are visible to the southern hemisphere observer looking towards the northern horizon in the different months of the year.

When the southern hemisphere observer looks north, to the point where the celestial equator reaches its maximum height, the regions of the celestial sphere to the north of the equator are below this line of reference (see Figure 5.1). The celestial north pole is inaccessible, as it is below the horizon. For the same

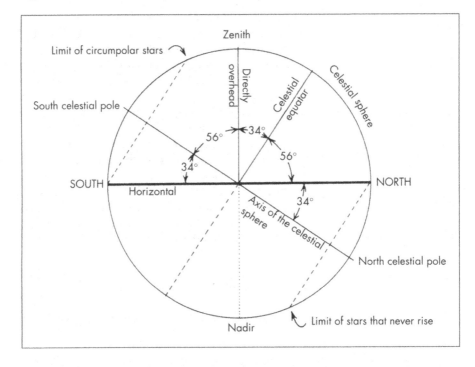

Figure 5.1. Conditions needed for certain stars to be circumpolar and others to never rise for an observer situated at latitude 34° south (34° S). This latitude corresponds approximately to that of the Cape of Good Hope, at the southern tip of Africa. Notice the angle made by the celestial equator with the horizontal plane of the place. For places in the Earth's southern hemisphere the celestial equator reaches its maximum height on the north side of the horizon. Using this diagram it is possible to draw conclusions about places of different latitude in the southern hemisphere, substituting 34° for the angle required and making any other necessary alterations.

reason, observers in this part of the world can never see stars less than 34° from the celestial north pole.

Seen from the southern hemisphere the constellations visible on the north side of the horizon seem upside down, in relation to their appearance in the northern hemisphere. The constellation of Orion (the Hunter) is a perfect illustration of this. The stars *Rigel* and *Saiph* are on the upper side of the constellation, and *Betelgeuse* and *Bellatrix* appear on the bottom side.

Let us have a look at some of the more curious aspects of the southern hemisphere sky, using as an example the conditions as they appear in Figure 5.1

and comparing them to Figures 3.7, 3.8, 3.9 and 3.10, which refer to northern hemisphere observation.

When the southern hemisphere observer looks north, the stars appear to move, as the hours pass, from right to left. These stars reach their maximum height on the north side of the horizon.

The Moon and the Sun reach their maximum height on the north side of the horizon and go on their apparent daily walks from right to left, which may be disconcerting to the northern hemisphere reader. There is no mystery about this. Looking north, the *east* is to your right and the *west* is to your left.

Unlike what we are used to in the northern hemisphere, the Moon is waxing when it looks like a letter C, and waning when it looks like a D.

Looking south the circumpolar stars wheel clockwise round the celestial south pole.

Appendix 9 shows approximate angular distances between some bright stars visible to the southern hemisphere observer. These stars are mentioned on Identification Maps of the corresponding regions of the sky. Appendix 7 shows the latitude of some cities in the southern hemisphere and close to the equator in English speaking countries.

5.3 Stars that Cannot Be Seen from Latitudes around 40° N

As mentioned earlier we will consider locations between latitudes 37° N and 42° N. The Central Region lies approximately at 40° N, the average latitude that can be used, more or less safely, by people living in these regions.

Therefore, we can say that for people living in places around latitude 40° N, the areas of the sky below 40° from the celestial south pole are invisible.

Identification Map M8 shows within the circle the stars that are not visible from parts of the world at approximately 40°N latitude. These stars and constellations of the celestial southern hemisphere are less than 40° from the celestial south pole, i.e. the part of the sky that is more than 50° south of the celestial equator.

In Map M8, the stars situated at less than 40° from the celestial south pole are those that are always above the horizon, in parts of the world with a latitude of 40° S, as shown in Figure 5.1. So stars that for observers at 40° N never rise are the same as those that never set for observers in places of latitude 40° S, i.e. circumpolar for these places. Equally, stars that are inaccessible to the observer at latitude 40° S, as they are permanently below the horizon, are the same as the circumpolar stars at 40° N, i.e. the circumpolar stars in parts of the world at latitude 40° N.

40° S is the latitude of Melbourne, Australia, as well as various places in Argentina, Chile and New Zealand. To make the maps easier to read, there is an overlap between the outer edge of Map M8 and the bottom parts of Maps M2 to M7. Regions between 50° and 55° south of the celestial equator appear in more than one map, so that the reader does not get lost when moving from one map to the next. Note that for observers in the southern hemisphere, Maps M2 to M7 should be turned upside down to match what is actually in the sky. For south hemisphere observers, the names of the months indicate the correspondence with the north direction of the horizon at the times shown.

5.4 Characteristics of Constellations Close to the Celestial South Pole

The constellations close to the celestial south pole were established around the beginning of the seventeenth century. In most cases these were not named after legendary figures. The first atlas of the stars to include constellations close to the celestial south pole was by a German lawyer called Johann Bayer (1572–1625). Published in 1603, this atlas was also the first to encompass the whole of the firmament. As far as the stars visible from Europe were concerned, Bayer based his work on information given to him by various navigators. He was responsible for introducing 12 new constellations: Pavo the peacock, Phoenix, Tucana the toucan, Grus the crane, Apus the bird of paradise,

Chamaeleon, Dorado the goldfish, Indus the Indian, Hydrus the lesser water snake, Triangulum Australe the Southern Triangle, Musca the fly and Volans the flying fish.

More recently, in the mid-eighteenth century, the French astronomer Nicolas Louis de Lacaille (1713–1762) introduced a further 14 constellations, which he named after scientific instruments or fine arts. They were Antlia the air pump, Caelum the chisel, Pyxis the compass, Fornax the furnace, Horologium the pendulum clock, Mensa the table, Microscopium, Telescopium, Norma the level, Octans, Circinus the drafting compass, Reticulum, Pictor the painter's easel and Sculptor. These constellations are very faint and can only be seen from dark places and even then with much difficulty. Lacaille also "dismembered" the old constellation of the ship (Argo Navis), which had been considered too large, and divided it into Puppis the stern, Vela the sails, Carina the keel and Circinus the drafting compass.

Various constellations in this region are relatively small, piled up in the sky, which makes it difficult to recognise them. In some cases they were created only to fill the gaps between other constellations, and their very faint stars are hard to detect unaided. These constellations do not stand out like those in the celestial northern hemisphere, so some are more difficult to identify. Appendix 5 lists the degree of difficulty in recognising constellations.

5.5 The Southern Cross: A Good Starting Point

Observers in the Earth's northern hemisphere use Ursa Major as their starting point for identifying stars and constellations. Similarly people in the southern hemisphere use the Southern Cross, the most famous constellation in the celestial southern hemisphere. In this case, the constellation is well named. The configuration of its four stars looks reasonably like a cross.

The Southern Cross is the smallest constellation in the entire sky. Its larger arm measures only 6°, between

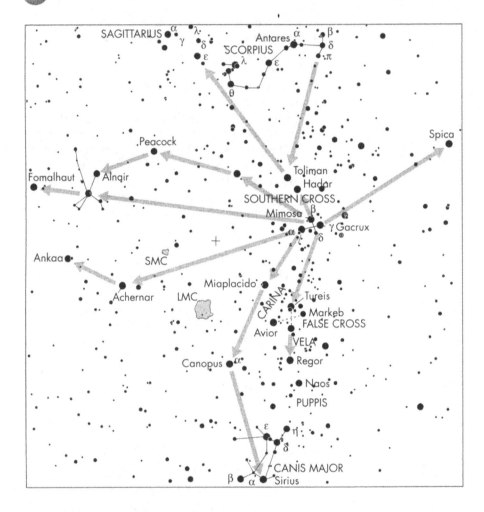

the stars Gacrux and Acrux. The smaller arm is a mere 4°, less than the width of two fingers. Three of its stars are very bright and the fourth is clearly visible, although the rest are extremely faint. From the Southern Cross, alignments can be set up leading to the brightest stars in the region. The Cross never sets in places south of latitude 33° S. The four stars that make up the shape of the cross are visible from the northern hemisphere, but only below 27° N, looking south on the horizon (22° if you want the stars to rise a little above the horizon).

This constellation is low, close to the southern horizon, at the beginning of the night in October and November (Map M8). It is at its highest after nightfall in May and June. If you wait until the Southern Cross

Figure 5.2.

reaches its highest point and its bigger arm is almost vertical, extending it to the horizon approximately enables you to find south.

Figure 5.2 can be turned so that it corresponds to the sky. You can change its position, so that The Southern Cross on the map looks like what you can see in the sky. Identification Map M8 enables you to do this, bearing in mind the month and time of observation. Moving Figure 5.2 you will find some of the alignments indicated and you can immediately put them into practice. The reader will see that even without measuring angles, the stars indicated here are easy to find – there is no way you can go wrong. Having said that, not all of the stars shown are visible on one occasion.

A further useful resource is the constellation Centaurus and its two brightest stars Toliman and Hadar. Toliman is also known as Alpha Centauri, Rigil Kentaurus and the Centaur's foot. Hadar is also called Agena.

The celestial south pole, marked with a + on Figure 5.2, can be located by extending the cross of The Southern Cross, by five times, in the direction Gacrux (γ Crucis) to Acrux (α Crucis). There is no particularly bright star at the celestial south pole. It is marked by a very faint star, magnitude 5.5, which is difficult to see with the naked eye, situated about 1° from the pole. It is also possible to find the celestial south pole by extending the larger arm of the cross as far as the star Achernar, the brightest star in Eridanus. The celestial south pole is halfway between the tip of the cross and Achernar. This star is especially bright and is surrounded by other fainter stars, which makes it easier to find.

Extending the larger arm of the Southern Cross the other way, from Gacrux to Acrux, you will see, slightly off the track, Spica, the brightest star in Virgo. The distance from Gacrux to Spica is around 40°, a little less than two hand spans. Following the direction defined by the stars Mimosa (β Crucis) and Acrux (α Crucis) of the Southern Cross, you reach first Miaplacido and Canopus, the brightest stars in Carina, the keel. Eventually you arrive at Sirius (α Canis Majoris).

Identifying the Southern Cross can be somewhat confusing, so it is essential that you make sure you have recognised it correctly. Indeed two of the stars in Carina, Avior and Tureis, also very bright, along with two stars in Vela, one of which is Markeb, make the shape of a cross. This shape, not a constellation, is known as the False Cross and the uninitiated may mistake this for the Southern Cross. The False Cross appears in Figure 5.2, on Map M8 and in Figure 6.5. Let us look at some of the differences between the two:

1. The Southern Cross is smaller than the False Cross.

2. There are two bright stars near to both crosses, more or less in line with their smaller arms, Regor (γ Velorum) and Naos (ξ Puppis), near the False Cross, and Toliman (α Centauri) and Hadar (β Centauri) near the real Southern Cross. The latter two stars are brighter and lie between The Southern Cross and Scorpius. Toliman is especially bright, only surpassed by Sirius (α Canis Majoris) and Canopus (α Carinae).

3. If there is still any doubt, Toliman and Hadar can be found by extending the line made by three stars of the Scorpion's head. Furthermore, the alignment of Hadar to Toliman points towards Sagittarius, a highly unusual constellation.

The alignment Gacrux to Mimosa leads you to Atria, the brightest star in Triangulum Australe. Moving on, you arrive at Peacock, the main star of Pavo, the Peacock. Moving around the curve of the Scorpion, you eventually reach the two most notable stars in Grus, the Crane. The brighter of the two is Alnair. There is no way you will get lost here. The stars indicated are by some distance the brightest in these areas of the sky, and stand out. Once you have identified these notable stars, you get some notion of the location of the various constellations, and can now move on to more detailed observation using Map M8.

Do not forget that there are other, fainter constellations "snuggling up" in the holes between some of the more notable constellations. Appendices 5 and 6 give useful ideas on the visibility and location of all the constellations.

The abbreviations LMC and SMC refer to Large Magellanic Cloud and Small Magellanic Cloud, two easily observable galaxies described in the text accompanying Map M8.

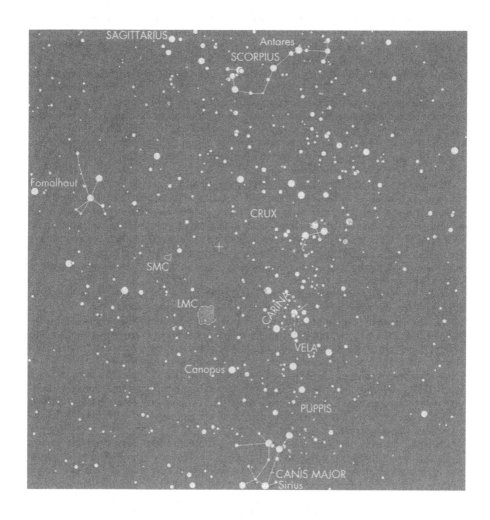

Identification Map 8

Circumpolar stars for latitude 40° S. Starting point: the Southern Cross

Note: Within its circumference, this map shows stars that are not visible from Europe or most of the USA. It can only be used by observers in the southern hemisphere, at latitudes close to 40° S. The map has only curiosity value for those in Europe. It can be useful to people in the northern hemisphere but considerably further south than Europe or most of the USA, although not as a map of circumpolar stars. But simultaneous observation of all the stars shown in the circle is only possible in places where the latitude is at least 40° S.

In order to use this map, the observer should face south, holding the map almost vertical in front. The map should be held in such a way that the name of the month corresponding to the time of year of the observation is at the top, pointing upwards. Now you have the aspect of the sky at the beginning of the night, around 10 p.m. As the hours go by each night, the stars wheel round clockwise, at a rate of 15° per hour. The lines dividing the names of the months are spaced 30° apart. For each hour later, turn the map round 15° clockwise, as shown by the arrows, so that what you see on the map matches what is actually in the sky. For each hour before, turn the map 15° anticlockwise.

The part of the circular map at the bottom at any given time shows the stars in the vicinity of the southern horizon, provided the horizon is unobstructed. The top part shows the stars close to the zenith. Stars outside the circumference are also shown here, as well as the names of the brightest stars, so as not to cut the sky abruptly and also to make it easier to connect this map with Maps M2 to M7. It is important to keep in mind the months when making this connection. For this reason, the map is square on the outside. It is useful in locations at latitude 30° S to 50° S, although for other latitudes, the circumpolar stars are not going to correspond exactly to what is indicated here.

Extending the larger arm of the Southern Cross (Crux), from Gacrux to Acrux, you eventually reach Achernar, the brightest star in Eridanus. It is quite far away, some 60°.

The smaller arm of The Southern Cross, extended towards Mimosa, allows you to identify the two brightest stars in Centaurus. Toliman, also known as Alpha Centauri, is the brightest; the other is Hadar, also called Agena (β Centauri). Toliman is only 4.3 light years away from us. It is a triple star, one of the components of which, Proxima Centauri, is slightly closer. After the Sun, it is in fact the closest to us. The second brightest star in Centaurus, Hadar (magnitude 0.6), is much further away, 520 light years. A line drawn straight across the sky from Toliman to Hadar takes you to The Southern Cross. Looking at Hadar and Toliman and remembering that one is 120 times further away from us than the other, even though they seem to be side by side, just goes to show how looks can be deceptive.

Following the direction from the Southern Cross and beyond the two brightest stars in Centaurus, Toliman and Hadar, takes you to the Scorpion's tail (see Figure 5.2). The configuration of the stars in this constellation actually look something like the shape of a Scorpion. In the position of the heart of the Scorpion is a very bright red star called Antares (α Scorpii).

Taking a line from Mimosa to Acrux (both in the Southern Cross), you reach Miaplacidus and then Canopus, the two brightest stars in Carina, the keel. Canopus is the brighter of the two and the second brightest in the sky, only surpassed by Sirius (α Canis Majoris). Between Miaplacidus and Canopus lies Volans, the Flying Fish, whose stars are not very bright.

The larger arm of the Southern Cross, from Gacrux to Acrux, points to the constellation of Musca, the fly, and then on to Chamaeleon, which only has faint stars. Between Toliman, the brightest star in Centaurus, and Atria, the brightest star in Triangulum Australe (the Southern Triangle), is the constellation Pyxis, the compass, which is extremely faint, but certainly reminiscent of the instrument after which it is named.

The stars Avior (ε Carinae) and Tureis (ι Carinae), along with Markeb (α Pegasi) and other bright stars in Vela make a shape like a cross in the sky. This is known as the False Cross, as mentioned in the notes on Figure 5.2.

Following other alignments that you may make for yourself using the maps it is possible to find further stars and constellations.

A line from Canopus (α Carinae) to Alnair (Grus, the crane) takes you eventually to the Large Magellanic Cloud (LMC) and the Small Magellanic Cloud (SMC). These are two small galaxies that are the closest to Earth and are satellites of our Galaxy. The brighter and larger of the two is the Large Magellanic Cloud, 180,000 light years away and situated in the constellation of Dorado the goldfish. To the naked eye it has an apparent diameter of 6° and is visible even with moonlight. Seen through binoculars on a dark night it is really spectacular. The Small Magellanic Cloud is less bright and of course smaller. It is located in the constellation of Tucana, the toucan, and is 230,000 light years away. It is easy to see with the naked eye on a dark night, and has a diameter of around 3.5°. It is more interesting when you look at it through binoculars.

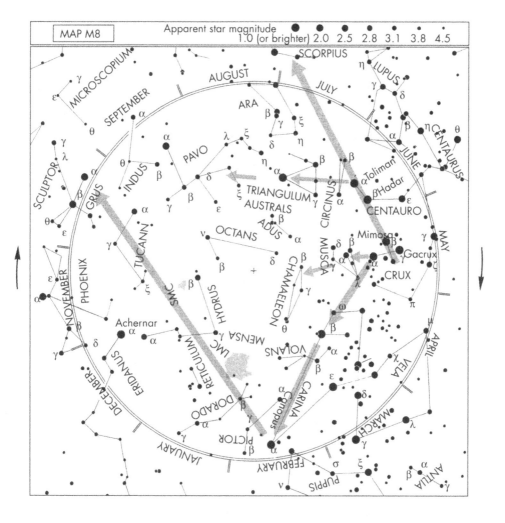

CHAPTER SIX

The Milky Way

As mentioned earlier, all of the stars we see unaided, or even with binoculars or small telescopes, belong to our Galaxy, usually known as *The Milky Way*. It is made up of one hundred thousand million (100,000,000,000) stars, some grouped together in clusters, and interstellar matter, dust and gas. All of the components of our Galaxy move around its nucleus. The stars are mostly spread out along spiral arms.

It is like one giant merry-go-round (see Figure 6.1). Our Galaxy measures around 100,000 light years in diameter and 12,000 light years at its maximum thickness. It is very widely spread out, rather like a colossal starry fried egg, as you can see in the diagram. The stars furthest from the galactic nucleus take longer to complete one lap than those that are nearer. Our Sun is around 28,000 light years from the centre and makes its giant orbit around the galactic centre at the fantastic speed of 250 km per second. Even so, it takes 220 million years to complete one lap!

The stars of the Galaxy are at various different stages of their lives: some just born, at the beginning of their long formation process, others more mature and approaching the ends of their lives. Some end their lives with a massive explosion, sending most of their matter into space. This matter forms giant clouds, which then combine to form new generations of stars.

The *star clusters* comes in two types. Some are less compact and situated in the galactic plain. These are made up of dozens or even hundreds of young stars and are known as open clusters. The easiest of these clusters to observe is that of The Pleiades, visible to the

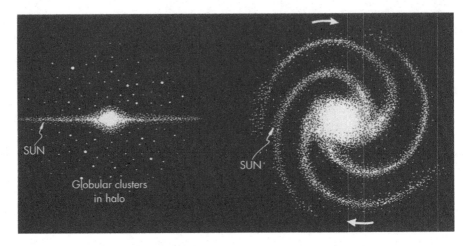

Figure 6.1. Schematic representation of our Galaxy. On the left, as seen in profile. On the right, as seen from the front, or from the top, if you prefer. Given the insignificance of the dimensions of the Solar System compared to the Galaxy, the position of the Sun as shown in this diagram also represents the position of our planetary system and, consequently, our position. The arrows indicate the direction of the Galaxy's rotation.

naked eye in the constellation Taurus (see Figure 4.7). Others are more compact and are known as *globular clusters*. These can contain hundreds of thousands of stars and are usually very old. They rotate around the galactic nucleus, following angled orbits, and form an approximately spherical halo around the Galaxy. There are more than 100 of these, but none is clearly visible to the naked eye.

There are indications that some stars, at a stable stage of their lives, may have planets orbiting them, as happens with our Sun.

Many stars in the Galaxy have accompanying stars, forming groups of two, three or more. These are referred to as multiple stars (double stars, triple stars and so on). In a system of multiple stars the component stars orbit a common centre of gravity.

We now know of thousands of millions of galaxies, of various shapes and sizes, some bigger and some smaller than ours. These are grouped together as "galaxy clusters". Usually our Galaxy is written with a capital "G", while other galaxies are spelt with a lower case "g".

On average, galaxies are an enormous distance away from each other. The two nearest are the relatively small Large Magellanic Cloud and Small Magellanic Cloud (see Map M8), named after the Portuguese navigator Magalhães. They are 230,000 and 180,000 light years away respectively and cannot be seen from countries such as Spain, central Europe and central America. The great galaxy of Andromeda, also one of the nearest, is detectable to the naked eye (see Figure 4.8) and is at a distance of 2,900,000 light years away, or 29 Milky Way "diameters". In proportion to their dimensions the galaxies are relatively close together.

Our Galaxy is a well spread-out structure and we are on its galactic plane. Consequently, provided your viewing location is dark enough, you should be able to see a milky strip crossing the night sky, which is not especially bright. This strip is commonly called the *Milky Way*, and is a vision of our Galaxy in profile around us (see Figure 6.2). Its width is irregular and its luminous aspect owes itself to the great concentration of distant stars in that region of the sky. To the naked eye, you are aware of its general luminosity, although it is impossible to distinguish between stars. With binoculars the strip "dissolves" into a multitude of remote suns, a magnificent spectacle.

The centre of the Galaxy is situated in the direction of the constellation of Sagittarius. Even so, clouds of interstellar matter make it impossible to observe the galactic nucleus. The constellation Auriga, the charioteer, can

Figure 6.2. Explanation of why we on Earth can see the shape of a milky strip crossing the sky, even though we are in the spread-out Galaxy. Looking in directions A, B, C or D, the directions of the galactic plane, observers on Earth will see a greater concentration of stars than when they look in, say, directions E or F. In direction A, we are facing the Galaxy's nucleus. Direction E is perpendicular to the galactic plane, also shown in this diagram.

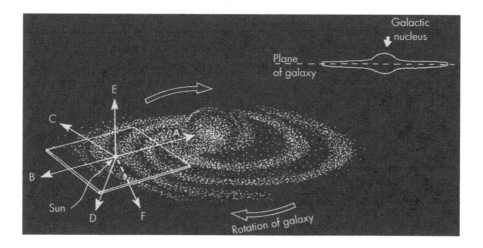

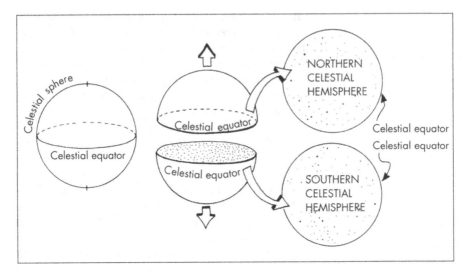

be found in the opposite direction. Looking towards Auriga, you are facing the periphery of the Galaxy.

The strip corresponding to the greatest density of stars in the Milky Way, and which marks the plane of our Galaxy, is not shown in the Identification Maps, or on the Celestial Charts (see next chapter). This is deliberate, so as to avoid confusion. The next two maps, Figures 6.4 and 6.5, together with Figure 6.3, are designed specifically for the reader to find the directions of the celestial sphere through which the strip passes. It also offers a panoramic view of the sky, given that these maps show simultaneously one celestial hemisphere each.

The name "Milky Way" is often used to refer to the white, milky strip that crosses the sky, in which there is the greatest density of stars, even though the other stars that live in the sky also belong to it, as mentioned earlier. The plane marked out by this strip approximately defines the galactic plane.

Figure 6.3. How the celestial sphere is flattened, for the purposes of the maps in Figures 6.4 (celestial northern hemisphere) and 6.5 (celestial southern hemisphere). As you can see in this diagram, the circumference of each of the maps is the celestial equator. In the centre of each of the maps is the relevant celestial pole.

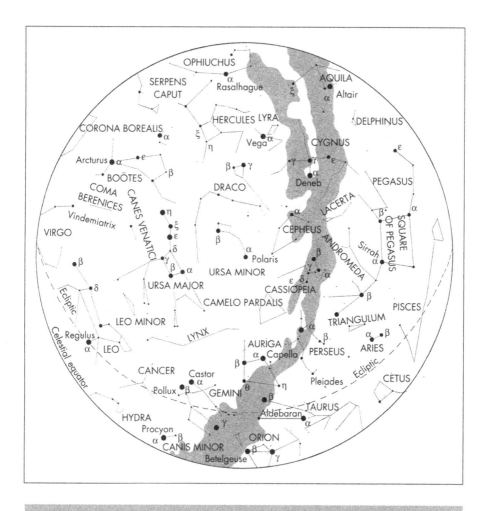

Figure 6.4. Representation of the celestial northern hemisphere and the Milky Way (grey strip). The circumference of this map is the celestial equator. The ecliptic is represented by longer broken lines. Some bright stars are also shown, by way of reference points. The constellations located in the direction of the celestial equator appear in part on this diagram and on the next.

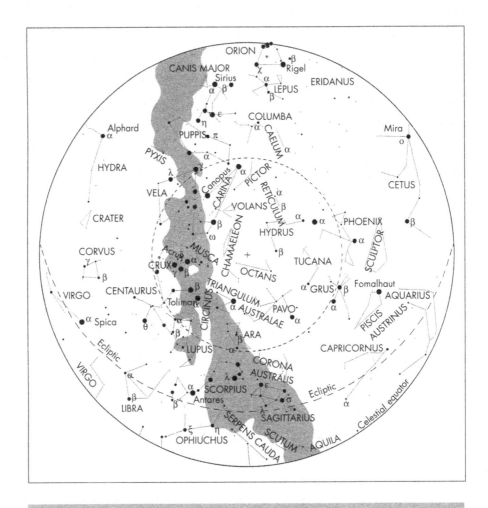

Figure 6.5. Representation of the celestial southern hemisphere and the Milky Way. The circumference of this map is the celestial equator. The smaller circumference, drawn with short broken lines, shows the limits of visibility for latitude 40° N (central USA), i.e. the stars within this circle are not visible from central USA. The ecliptic is represented by longer broken lines. The celestial south pole is marked with a +. The names of some bright stars are also in the diagram, to serve as a reference.

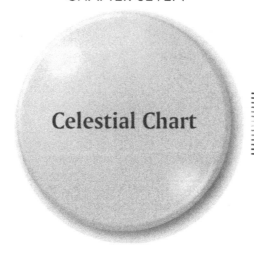

CHAPTER SEVEN

Celestial Chart

Identification Maps M1 to M8 give a number of indications enabling the interested reader to identify dozens of bright stars and most of the constellations. The arrows showing the alignments and extensions, however, make it difficult or even impossible to use them as maps of the sky. In order to overcome these problems, this chapter – essentially one Celestial Chart – includes these maps from a new perspective.

The maps of the Celestial Chart carry a letter C, along with a number corresponding to the Identification Map of the *same* region of the celestial sphere. For example, Map C3 shows the same part of the sky as Map M3. Each map takes up two pages; on the left-hand page, you will find the same map as the Identification Map, but without the arrows and alignments; the right-hand page contains a map of the same area *in positive*, i.e. stars in white on a black background. These positive maps do not contain lines to link the brightest stars in each constellation, nor do they include the names of stars or constellations (except for C1, which shows the names of some stars). The maps on the right give a more realistic representation of the night sky and are aimed at training the reader. They should be used in conjunction with the actual stars themselves.

If you are in any doubt, the left-hand pages will give you the support you need. You can also go back to the corresponding Identification Map, whenever you wish, and use the accompanying notes and the arrows showing the alignments and some geometric shapes. Appendix 4 gives information on all stars whose names appear on the maps.

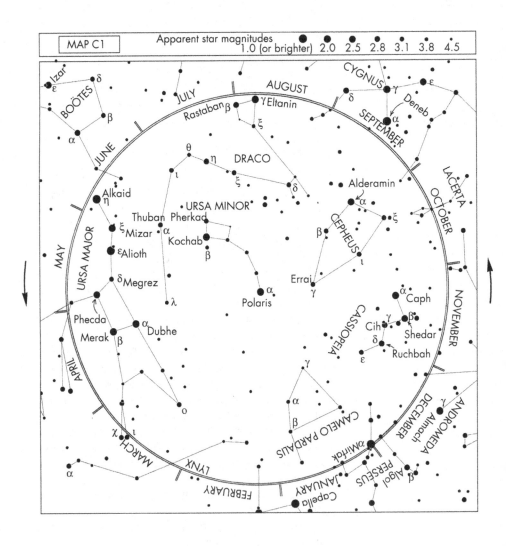

NOTE: Figure 6.4 shows a panoramic view of the celestial northern hemisphere.

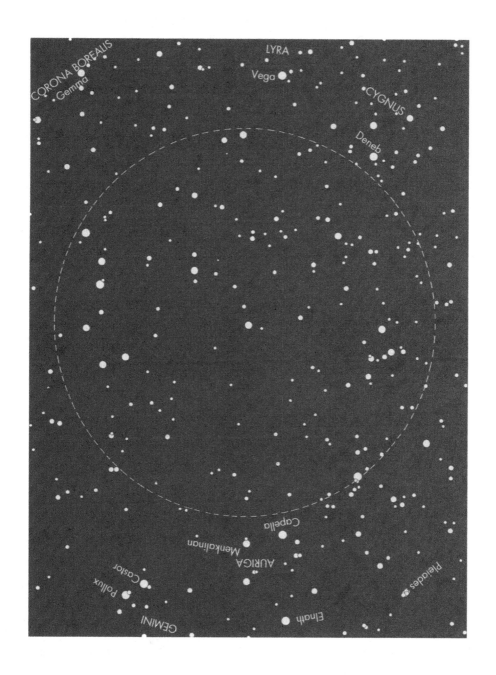

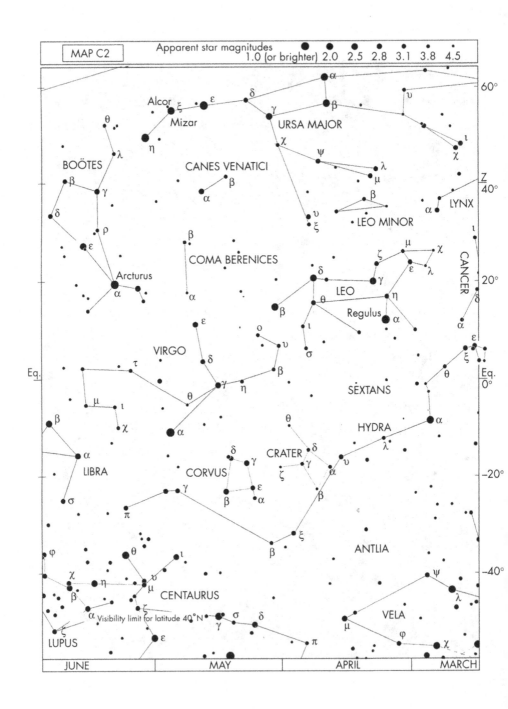

MAP C2

Apparent star magnitudes
1.0 (or brighter) 2.0 2.5 2.8 3.1 3.8 4.5

Alcor ξ ε δ α URSA MAJOR β υ ι χ 60°

Mizar γ

θ λ η χ ψ λ Z 40°

BOÖTES CANES VENATICI μ

β β α β LEO MINOR LYNX

γ υ α

δ ξ CANCER

ρ μ χ

β ζ ε λ 20°

COMA BERENICES δ γ δ

Arcturus α α θ LEO η α

ε Regulus α

VIRGO o ε δ

τ υ σ

Eq. δ β SEXTANS Eq.
0°

μ ι θ η ε

β θ ξ

χ α HYDRA α

α λ

σ δ γ CRATER δ

LIBRA CORVUS ζ γ υ

π γ ε β β

β α

ξ ANTLIA -20°

β

θ ψ

φ ι λ

χ η υ

β μ CENTAURUS VELA -40°

α Visibility limit for latitude 40°N ζ σ δ μ φ

LUPUS ζ γ χ

ε π

JUNE | MAY | APRIL | MARCH

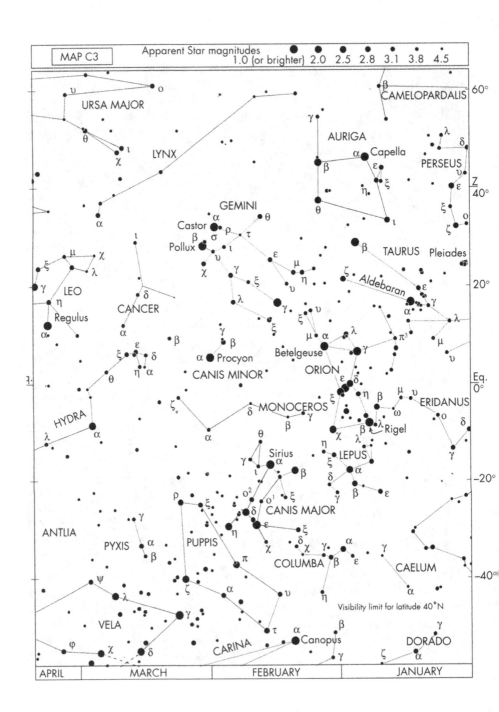

MAP C3

Apparent Star magnitudes
1.0 (or brighter) 2.0 2.5 2.8 3.1 3.8 4.5

60°

CAMELOPARDALIS

β

URSA MAJOR

υ   o

θ
χ   ι
LYNX

AURIGA

γ

α Capella
β
ε
ξ
η

λ
δ
PERSEUS
υ
ε

ξ
o
ζ

Z
40°

α

GEMINI

α   θ

Castor
β σ ρ τ
Pollux ι
υ
χ γ ξ
λ

ε
μ
η
υ

β TAURUS Pleiades

ζ
Aldebaran

ε
γ
α
λ

20°

ξ μ χ
γ
ξ   λ

θ
ι
LEO
η
Regulus

δ
CANCER

α

β

γ
β

γ
ξ

μ α
λ

υ
ξ

Betelgeuse γ
ORION

π³
μ
υ

α

ε
δ

Procyon
CANIS MINOR

HYDRA
λ

ε
ξ δ
η α
θ

α

ζ

δ
MONOCEROS
β γ

ξ
η β μ υ
ω
β λ Rigel
χ λ

ERIDANUS
o
δ

γ

Eq.
0°

-20°

ANTLIA

PYXIS

γ

α
β

ζ

ρ
ξ

PUPPIS

π

θ

Sirius
γ α
ι
β

o²
o¹ ξ
δ CANIS MAJOR
ε
χ δ χ γ
COLUMBA β ε

η β
ξ α
δ γ
ξ α
LEPUS
β
γ β ε

α

γ

CAELUM

α

-40°

VELA

ψ
λ

φ χ δ

γ

ζ
α

τ α Canopus
CARINA
γ

υ
η

β

Visibility limit for latitude 40°N

γ
DORADO
ζ α

APRIL   MARCH   FEBRUARY   JANUARY

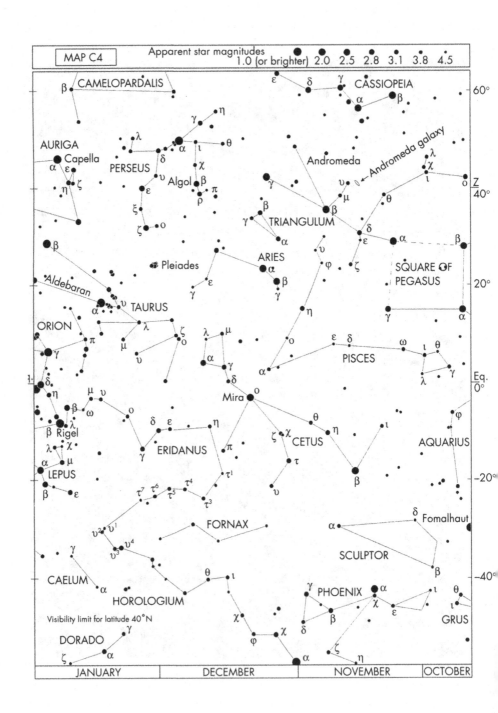

MAP C4

Apparent star magnitudes
1.0 (or brighter) 2.0 2.5 2.8 3.1 3.8 4.5

60°

CAMELOPARDALIS
β

CASSIOPEIA
ε δ γ
α β

AURIGA
Capella
α ε
η ζ

PERSEUS
γ η
λ
δ α ι θ
χ
υ Algol β
ξ π
ρ
ζ ο

Andromeda
Andromeda galaxy
λ χ
ι
υ θ ο
μ

Z
40°

β TRIANGULUM
γ β δ
α ε α β

SQUARE OF
PEGASUS

20°

Pleiades
Aldebaran
υ TAURUS
α λ

ARIES
α
β
γ

η

υ
φ ζ

γ α

ORION
π μ ζ
λ ο
υ

λ μ
α γ
δ

PISCES
ε δ ω ι θ
λ γ

Eq.
0°

γ
δ
η
μ υ
β ω ο
λ χ
α μ

Mira
ο

θ
ζ χ η CETUS
τ

ι
φ

AQUARIUS

β Rigel
LEPUS
β ε

δ ε η
γ ERIDANUS
π
τ¹

υ

β

−20°

τ⁷ τ⁶ τ⁴
τ⁵
τ³

FORNAX
α

δ Fomalhaut
β

υ² υ¹
υ³ υ⁴
γ

SCULPTOR

−40°

CAELUM
α

θ ι

γ
PHOENIX α
χ ε
β
δ

ι θ
ι
GRUS

HOROLOGIUM
Visibility limit for latitude 40°N

χ
φ χ
ζ
α η

DORADO
ζ α
γ

| JANUARY | DECEMBER | NOVEMBER | OCTOBER |

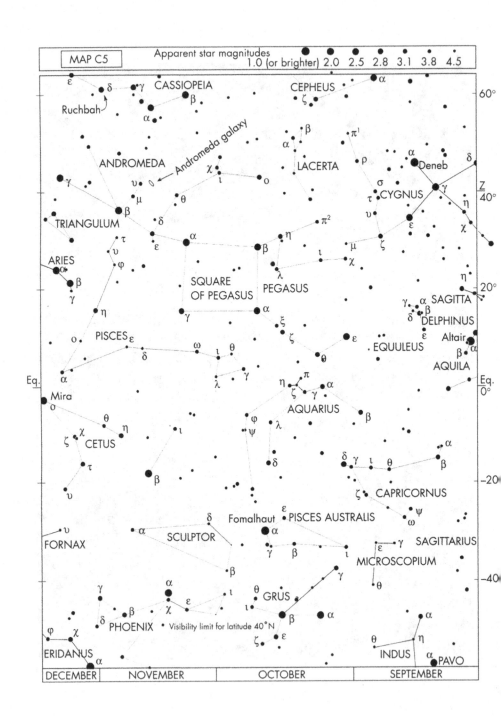

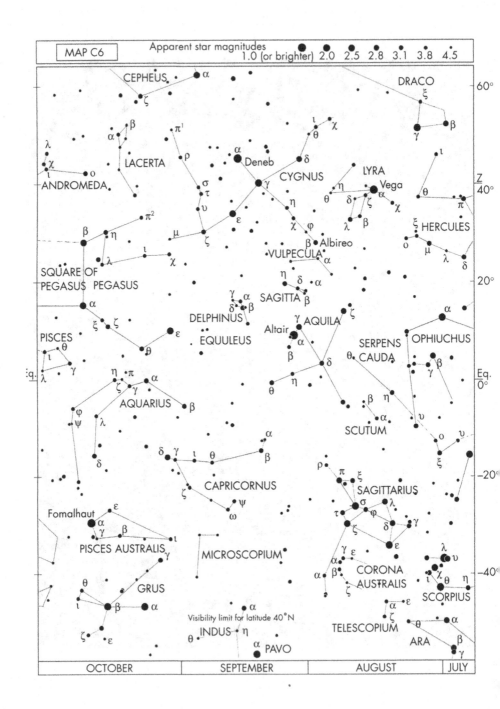

MAP C6

Apparent star magnitudes
1.0 (or brighter) 2.0 2.5 2.8 3.1 3.8 4.5

OCTOBER SEPTEMBER AUGUST JULY

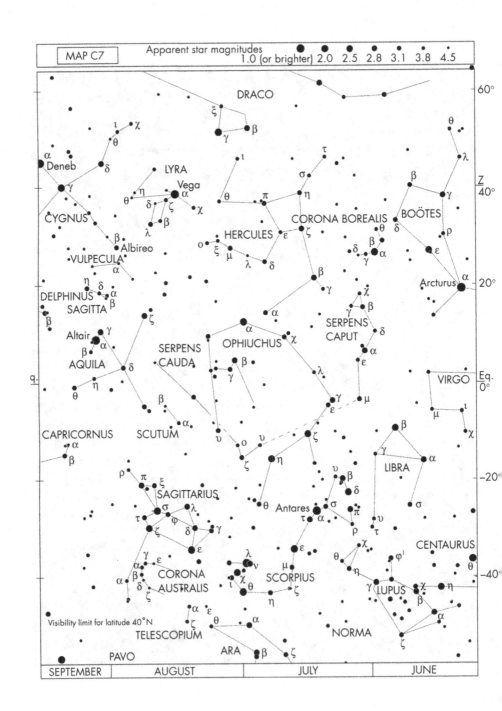

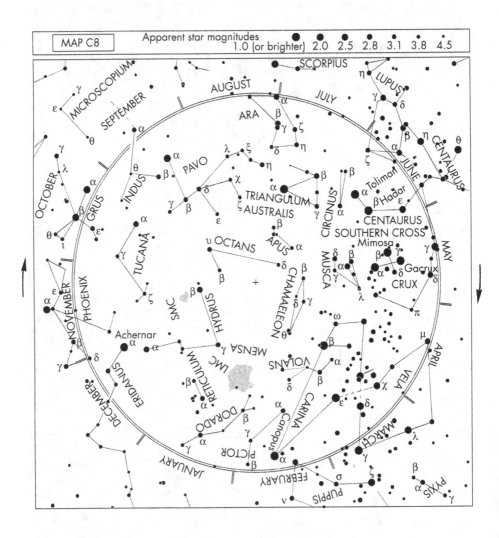

NOTE:
1. The names of the months September and October are further outside, so as not to cover up the stars shown in that part of the sky.
2. Figure 6.5 shows a panoramic view of the celestial southern hemisphere. See also Figure 5.2.

The previous chapter (Figures 6.4 and 6.5) will also help you to locate the strip containing the most populous areas of the Milky Way. Once you know how to identify a number of constellations, it is well worth looking through this region with binoculars, preferably in a place where the sky is very dark.

If you lie down on a comfortable surface you will be able to see as far as the zenith without having to lean your head too much. This is the recommended position for observing the regions of the sky close to the zenith, unaided or with small binoculars. Usually, though, it is better to mount your binoculars on a tripod. There are special accessories for this purpose available in photographic shops. This way you avoid your hands shaking, a problem afflicting everyone to a greater or lesser degree, and your arms getting tired.

CHAPTER EIGHT

The Sky and the Four Seasons of the Year

The various maps in this book show the procedures to follow for confident recognition of the sky. However, it may sometimes be useful to have a panoramic representation of the whole of the sky visible at any given moment. To make it easier to locate the main constellations, each of the eight maps in this chapter shows the overall aspect of the sky at around 9.30 p.m. (winter time) and 10.30 p.m. (summer time) in the middle of each of the seasons of the year for the northern and southern hemispheres. They also show the sky's appearance at other times and dates, as indicated. The dates in *italics* are outside the seasons indicated; those in **bold** correspond to the times above.

The seasons of the year alternate between the northern and southern hemispheres, as in Table 8.1.

The "or" in this table has a simple explanation. Due to the need to maintain the calendar in keeping with the seasons of the year, normal years and leap years had to be taken into account. The system therefore accumulates a slight discrepancy of 6 hours per year, which are in part corrected every four years. This accumulation leads to a variation of one day on the dates

Table 8.1. Alternation of the seasons in the northern and southern hemispheres

Approximate date	Northern hemisphere	Southern hemisphere
20 or 21 March	Start of spring	Start of autumn
20 or 21 June	Start of summer	Start of winter
22 or 23 September	Start of autumn	Start of spring
21 or 22 December	Start of winter	Start of summer

mentioned, although these remain within the limits indicated. Each season begins when the previous one finishes, of course.

Time differences have been taken into account, i.e. summer time and winter time, the dates in summer time marked with an asterisk*. These maps show the sky as it is seen by an observer lying on the ground looking up and finding the zenith in the corresponding position at the centre of each map. Only the most apparent constellations and the most notable stars are included, since the aim of these maps is locating secure references in the sky. The Identification Maps and the Celestial Charts give far more detail and explanation regarding the area of the sky you wish to look at. Using these maps is easy.

Choose the appropriate map, according to the season of the year and the time of day. Hold the map in a *horizontal* position, above your head, and turn the part of the map showing NORTH in the corresponding direction on the horizon. In this way, the map is correctly positioned and shows the main constellations, as well as the names of some reference stars. Keeping the map in position, the observer can look in any direction on the horizon or towards the zenith. Alternatively you can hold the map *vertically*, so that the name corresponding to the direction on the horizon in which you wish to look is *facing downwards*. So, if you want to look to the west, face west and turn the map round, so that the word west is at the bottom. It is now possible to see the positions of the main constellations, no matter what direction you look in.

Some stars outside the horizon are shown on this map, in the west and in the east. These are stars that are going to rise soon or those that have just set. The labels have been set out in such a way that they always appear on the right, regardless of the direction in which the observer is looking, making them easier to read.

The first four maps refer to the sky observable in temperate regions of the northern hemisphere and were specifically designed for the average latitude of 40°N, corresponding to countries such as Spain, Italy and Central USA. The last four show the sky seen in temperate regions of the southern hemisphere, and were made for latitude 40° S. In the latter maps, the abbreviations SMC and LMC refer to the Small Magellanic Cloud and the Large Magellanic Cloud. For the southern hemisphere, summer time and winter time have also been included. The titles of the south-

Northern hemisphere

Spring
(21 March to 21 June approx.)

Appearance of the sky on 1 May at approx. 10.30 p.m.
Also observable on these dates and at these times:

30 January at 3.30 a.m.   15 April at 11.30 p.m.*
14 February at 2.30 a.m.   **1 May at 10.30 p.m.***
1 March at 1.30 a.m.   15 May at 9.30 p.m.*
15 March at 12.30 a.m.   30 May at 8.30 p.m.*
1 April at 12.30 a.m.*

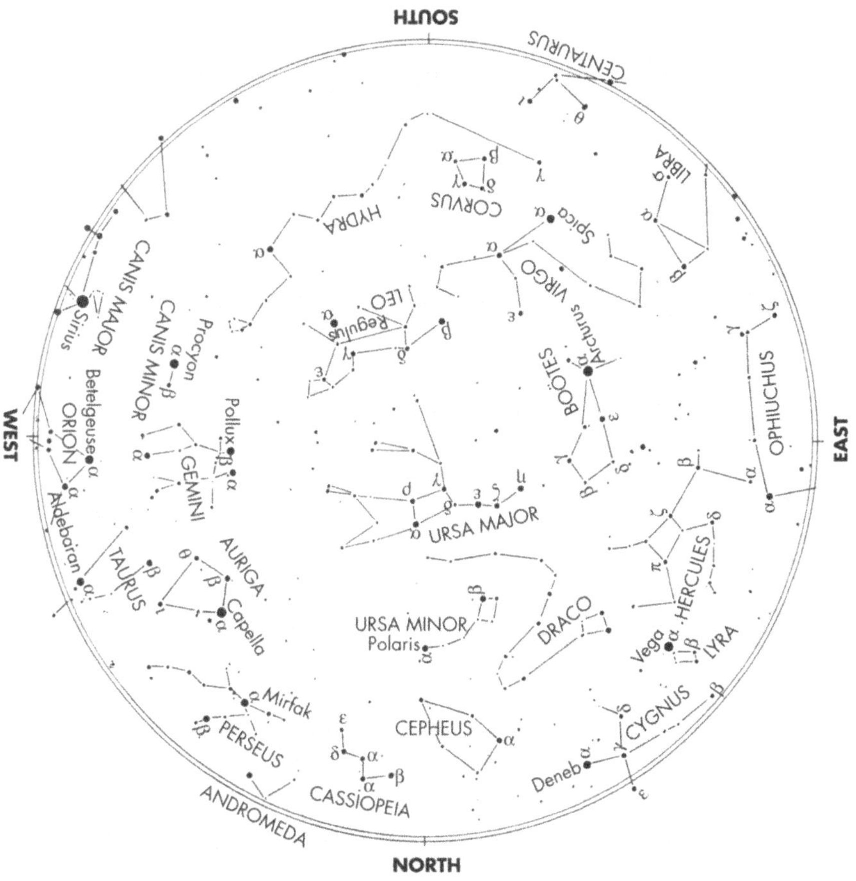

Figure 8.1.

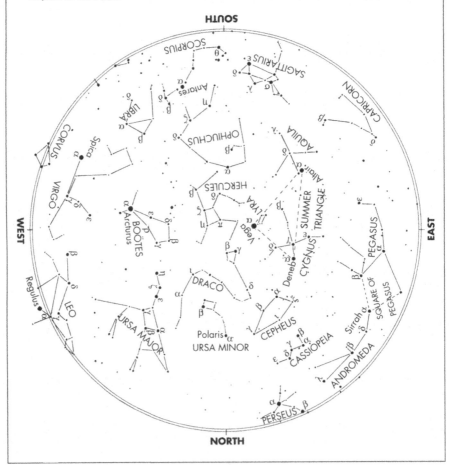

Northern hemisphere   **Summer**
(21 June to 23 September approx.)

Appearance of the sky on 1 August at approx. 10.30 p.m.
Also observable on these dates and at these times:

15 April at 5.30 a.m. *
1 May at 4.30 a.m. *
15 May at 3.30 a.m. *
1 June at 2.30 a.m. *
15 June at 1.30 a.m. *

1 July at 12.30 a.m. *
15 July at 11.30 p.m. *
1 August at 10.30 p.m.* *
15 August at 9.30 p.m. *

Figure 8.2.

Northern hemisphere   Autumn
(23 September to 22 October approx.)

Appearance of the sky on 1 November at approx. 9.30 p.m.
Also observable on these dates and at these times:

15 August at 3.30 a.m.*
1 September at 2.30 a.m.*
15 September at 1.30 a.m.*
1 October at 12.30 a.m.*

15 October at 11.30 p.m.*
1 November at 9.30 p.m.
15 November at 8.30 p.m.
30 November at 7.30 p.m.

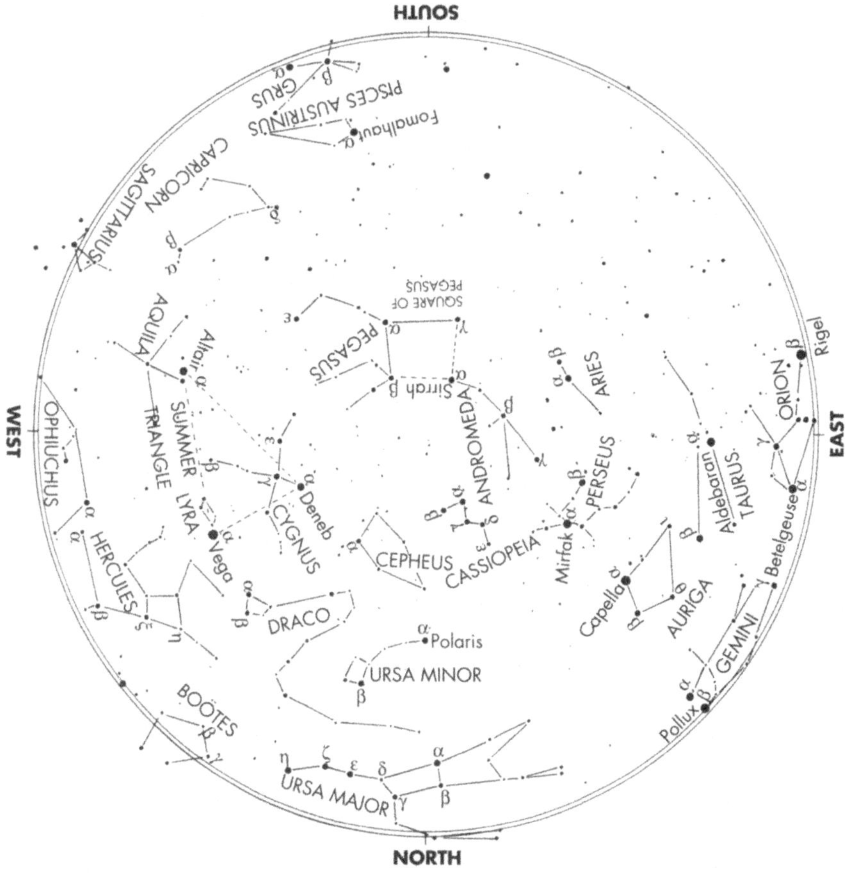

Figure 8.3.

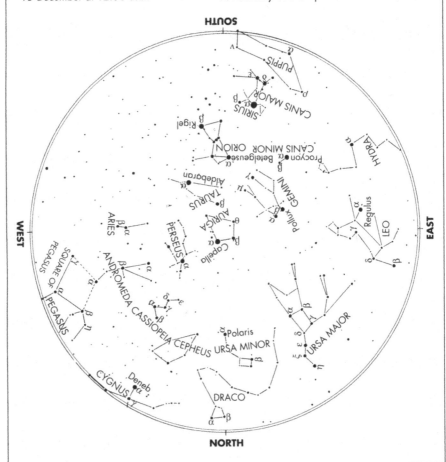

Figure 8.4.

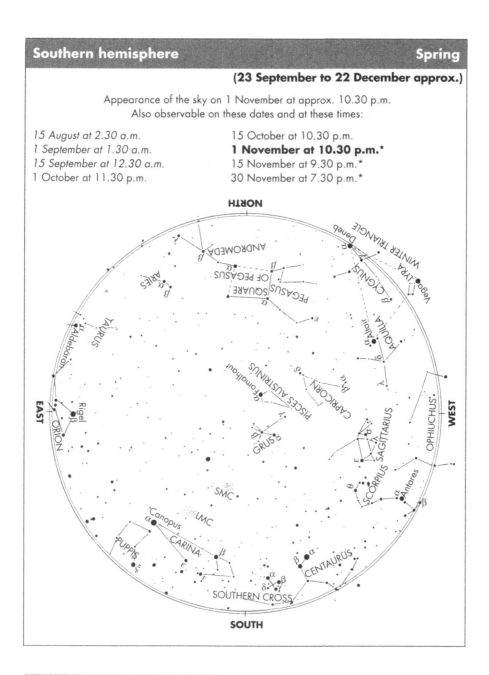

Southern hemisphere **Spring**

(23 September to 22 December approx.)

Appearance of the sky on 1 November at approx. 10.30 p.m.
Also observable on these dates and at these times:

15 August at 2.30 a.m.
1 September at 1.30 a.m.
15 September at 12.30 a.m.
1 October at 11.30 p.m.

15 October at 10.30 p.m.
1 November at 10.30 p.m.*
15 November at 9.30 p.m.*
30 November at 7.30 p.m.*

Figure 8.5.

Southern hemisphere   Summer

(23 December to 21 March approx.)

Appearance of the sky on 1 February at approx. 10.30 p.m.
Also observable on these dates and at these times:

1 November at 4.30 a.m. *
15 November at 3.30 a.m. *
1 December at 2.30 a.m. *
15 December at 1.30 a.m. *

1 January at 12.30 a.m. *
15 January at 11.30 p.m. *
**1 February at 10.30 p.m.* **
15 February at 9.30 p.m.

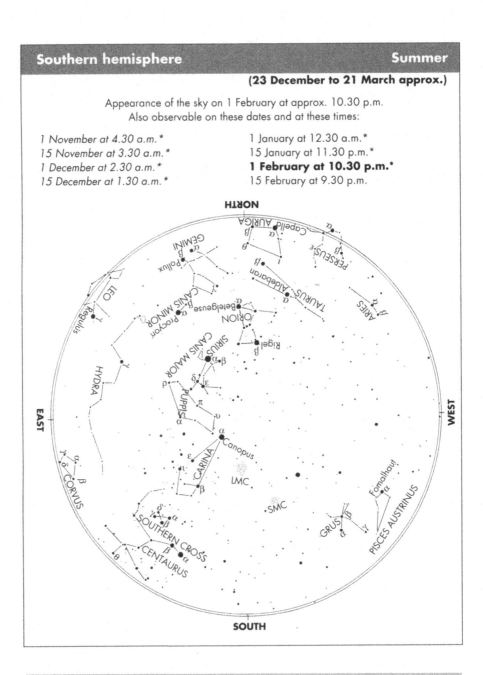

Figure 8.6.

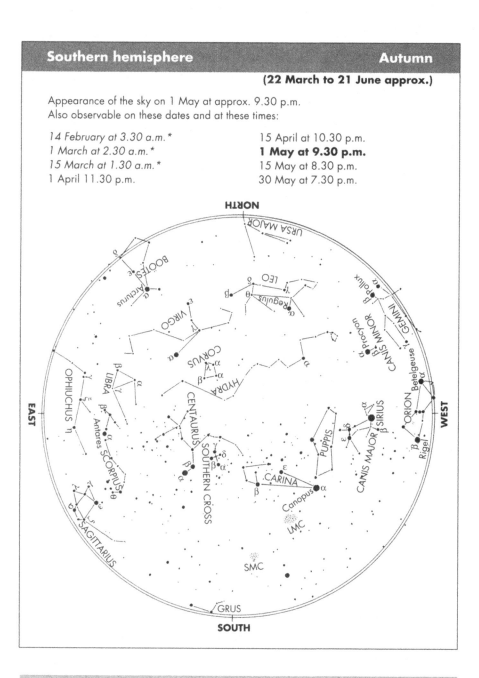

Southern hemisphere   **Autumn**

(22 March to 21 June approx.)

Appearance of the sky on 1 May at approx. 9.30 p.m.
Also observable on these dates and at these times:

14 February at 3.30 a.m. *	15 April at 10.30 p.m.
1 March at 2.30 a.m. *	**1 May at 9.30 p.m.**
15 March at 1.30 a.m. *	15 May at 8.30 p.m.
1 April 11.30 p.m.	30 May at 7.30 p.m.

Figure 8.7.

Southern hemisphere — Winter

(22 June to 23 September approx.)

Appearance of the sky on 1 August at approx. 9.30 p.m.
Also observable on these dates and at these times:

1 May at 3.30 a.m.
15 May at 2.30 a.m.
1 June at 1.30 a.m.
15 June at 12.30 a.m.

1 July at 11.30 p.m.
15 July at 10.30 p.m.
1 August at 9.30 p.m.
15 August at 8.30 p.m.

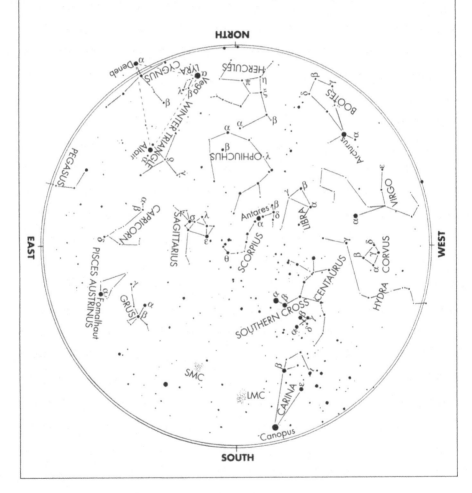

Figure 8.8.

ern hemisphere maps are in negative, i.e. white on a black background, in order to make the distinction clear between the two hemispheres.

These maps, just as the Identification Maps and the Celestial Charts, are valid throughout the year. As such they can be considered perpetual.

8.1 What To Do Next

By comparing the various maps in this book with what is actually in the sky in front of you and following the method of alignments and geometric shapes in the book, you have learnt what is needed to be able to identify stars and constellations.

Maybe you have created your own alignments and techniques for identifying stars and constellations. These are of course just as valid, as long as everything is identified correctly. You have also acquired habits that ought to stand you in good stead for more ambitious observations in the future. These are procedures followed by people who are well used to looking at the stars, people who know the pathways to knowledge of the stars.

You have discovered that the sky's appearance varies according to the time of year and time of night. You now know that at certain times of year, you cannot observe certain stars and that others are inaccessible due to your observation latitude. Following the indications in this book, you might well have come to the conclusion that the sky is not quite as difficult as you thought. From the knowledge and procedures you have learned in this book, you can widen your observations and use binoculars or even small telescopes. After all, the Universe is there for anyone who wants to observe it.

The first part, even if you were starting from scratch, is now out of the way. Now you should feel at ease with these procedures, having gained confidence and should be able to undertake new explorations. This was the main aim of *Navigating the Night Sky*.

Once you know the sky with the naked eye, using binoculars will prove extremely useful. Binoculars will never stop being of use in astronomical observation, even for people who own a telescope. I recommend a 7 × 50 model (sevenfold magnification, 50 mm diameter lenses), which is capable of encompassing a field of view of 7° in diameter, or 10 × 50, whose field of view is around 6°.

With binoculars, you can observe immense fields of stars in the plane of the Milky Way, various star clusters – globular or open – some nebulae and even some of the brighter and more remote galaxies. You will be amazed by the staggering array of stars offered by this accessible instrument. It is possible to recognise various aspects of the lunar relief and see the three biggest moons of Jupiter. Your binoculars should be mounted on a tripod. There are accessories you can buy at photographic shops to help with this.

With a small telescope, observation becomes even more interesting and diverse. In addition to all of the above, you can also explore craters, valleys and mountains on the surface of the Moon, observe the phases of Venus and Mercury, or appreciate the beauty of Saturn and its majestic rings, which accompany its biggest satellite, Titan. You will be able to observe Jupiter's

globe ringed by some of its satellites, check on the Sun's rotation, see many clusters, some nebulae and galaxies, and so on.

For these and other observations, you will need to know what is above the horizon at any given point in time and what is not. *You need to know the sky.* Only if you know the sky will you be able to know where to look and in which direction to point binoculars or a telescope. I hope that this book has opened up the pathway to the stars for you. You should now have a strong base from which to use more elaborate maps of the sky and to explore the celestial wonders.

I hope that you enjoy your observations. As I have said on a number of occasions, this book is supposed to be used, in the main, in front of the stars themselves. Think of them as your advisors and try to share this advice with others.

8.2 Constellations with which the Reader is Already Familiar

The reader should bear in mind that not all of the constellations are accessible from Europe or North America. Other stars may not be visible, as they may be below the horizon on the date and at the time of observation. For further information, please see the Identification Maps and Appendices 5 and 6. The following is an alphabetical list of the constellations that you can tick off once you have located them in the sky.

☐ Andromeda
☐ Antlia (Air Pump)
☐ Apus (Bird of Paradise)*
☐ Aquarius (Water Bearer)
☐ Aquila (Eagle)
☐ Ara (Altar)*
☐ Aries (Ram)
☐ Auriga (Charioteer)
☐ Boötes (Herdsman)
☐ Caelum (Chisel)
☐ Camelopardalis (Giraffe)

☐ Cancer (Crab)
☐ Canes Venatici
  (Hunting Dogs)
☐ Canis Major (Great Dog)
☐ Canis Minor (Little Dog)
☐ Capricornus (Sea Goat)
☐ Carina (Keel)*
☐ Cassiopeia
☐ Centaurus (Centaur)
☐ Cepheus
☐ Cetus (Sea Monster)

☐ Chamaeleon (Chameleon)
☐ Circinus (Drafting Compasses)
☐ Columba (Dove)
☐ Coma Berenices
  (Berenice's Hair)
☐ Corona Australis
  (Southern Crown)
☐ Corona Borealis
  (Northern Crown)
☐ Corvus (Crow)
☐ Crater (Cup)

- ☐ Crux (Southern Cross)*
- ☐ Cygnus (Swan)
- ☐ Delphinus (Dolphin)
- ☐ Dorado (Swordfish)*
- ☐ Draco (Dragon)
- ☐ Equuleus (Foal)
- ☐ Eridanus (River)
- ☐ Fornax (Furnace)
- ☐ Gemini (Twins)
- ☐ Grus (Crane)
- ☐ Hercules
- ☐ Horologium (Clock)
- ☐ Hydra (Sea Snake)
- ☐ Hydrus (Sea Serpent)*
- ☐ Indus (Indian)*
- ☐ Lacerta (Lizard)
- ☐ Leo (Lion)
- ☐ Leo Minor (Little Lion)
- ☐ Lepus (Hare)
- ☐ Libra (Scales)
- ☐ Lupus (Wolf)
- ☐ Lynx (Lynx)
- ☐ Lyra (Lyre)

- ☐ Mensa (Table)*
- ☐ Microscopium (Microscope)
- ☐ Monoceros (Unicorn)
- ☐ Musca (Fly)*
- ☐ Norma (Square)
- ☐ Octans (Octant)*
- ☐ Ophiuchus (Serpent Bearer)
- ☐ Orion
- ☐ Pavo (Peacock)*
- ☐ Pegasus (Winged Horse)
- ☐ Perseus
- ☐ Phoenix*
- ☐ Pictor (Easel)*
- ☐ Pisces (Fishes)
- ☐ Piscis Austrinus (Southern Fish)
- ☐ Puppis (Stern)
- ☐ Pyxis (Compass)*
- ☐ Reticulum (Reticle)*
- ☐ Sagitta (Arrow)
- ☐ Sagittarius (Archer)

- ☐ Scorpius (Scorpion)
- ☐ Sculptor (Sculptor)
- ☐ Scutum (Shield)
- ☐ Serpens (Serpent):
  Serpent Caput (Serpent's Head)
  Serpent Cauda (Serpent's Tail)
- ☐ Sextans (Sextant)
- ☐ Taurus (Bull)
- ☐ Telescopium (Telescope)
- ☐ Triangulum (Triangle)
- ☐ Triangulum Australe (Southern Triangle)*
- ☐ Tucana (Toucan)*
- ☐ Ursa Major (Great Bear)
- ☐ Ursa Minor (Little Bear)
- ☐ Vela (Sails)
- ☐ Virgo (Maiden)
- ☐ Volans (Flying Fish)*
- ☐ Vulpecula (Little Fox)

*Constellation not observable from latitudes close to 40° N. Also marked are constellations of which only a small part appears above the horizon in that locations.

Examples of Ancient Representations of Some Constellations

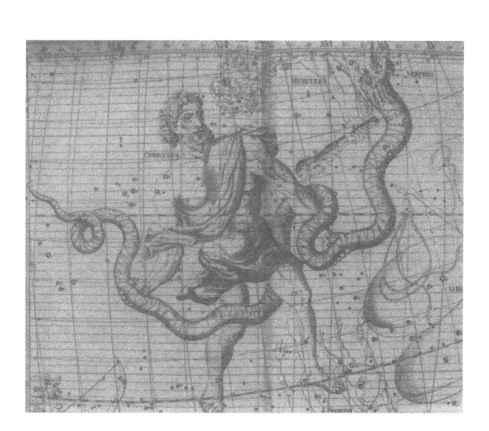

Figure A.1. Ophiuchus (the serpent bearer), according to the *Celestial Atlas* of John Flamsteed (1646–1719), published posthumously in 1725.

It is rare that a constellation is recognised because of any possible similarity between the configuration of its brightest stars and the shape after which the constellation was named. Only in very few constellations, especially the oldest, can one truly say that the resemblance is sufficiently clear. It is important to bear in mind that these shapes have nothing to do with modern astronomy. Out of curiosity, you may wish to know more about these figures and the legends associated with them, but you should not approach them with the intention of finding the resemblance in any definitive way. So it is worth a look at the ancient figures and their representations, the way in which they evolved and the reasons why they were ultimately abandoned. Mostly, the ancient names have been kept and this link to the past is a fascinating heritage that you can awaken and share with other people, when you look at the sky on a clear and calm night.

The first constellations were imagined, over 6000 years ago, by the Sumerians and then the Babylonians, who inhabited Mesopotamia, situated between the rivers Tigris and Euphrates in modern day Iraq. It is, however, impossible to establish exactly when man first started to "populate" the firmament with his imagination. This first systematisation of the sky was passed on to other peoples with whom the Babylonians came into contact and later arrived in Greece. There are explicit references to the constellations and to some stars in the books of various Greek authors, dating back to the eighth century BC. Using the celestial shapes, the Greeks, and then the Romans, made various adaptations and recreated some of their own legends in the sky, which have come down to us. In the year 143 BC, the Greek astronomer Hipparchus (190–120 BC) started to make the first catalogue of the stars, which future generations of astronomers could use to detect any alterations in the sky's appearance.

More than half of the 88 current constellations have their origins in the great systematisation of the science of celestial bodies made by the astronomer Claudius Ptolemy (85–160 AD), who lived and worked in Alexandria, Egypt. Finished in 150 AD, the work was monumental, containing 12 volumes. It became known for its later Arabic translation entitled *Almagest*, meaning The Greatest. In the work, which was in part a compilation of the knowledge acquired by previous generations of astronomers, Ptolemy included 48 constellations, although he did not invent them – they had already been imagined long before his time. He specified the positions of the stars within the figures imagined – certain stars marked the head, others the left hand, and so forth. Later, Arab astronomers adopted Ptolemy's constellation system, although they kept some designations from their own tradition. The first printed map of the stars appeared much later, in 1515, and was the work of Albrecht Dürer, a German painter and engraver. The remaining 40 constellations were subsequently added by successive astronomers in the sixteenth, seventeenth and eighteenth centuries, thus filling the gaps between Ptolemy's constellations and occupying regions around the celestial south pole, which were not observable either from Greece or Mesopotamia.

The oldest drawings of the figures associated with the constellations have a symbolic character. The stars were marked in relative positions, which did not match the appearance of the sky. Later the positions of the stars started to be marked with greater precision in richly coloured maps showing well-proportioned and realistic shapes. In some cases there was such a profusion of colour and detail that it was actually difficult to find the stars. Artistic considerations were paramount. Many of these maps were made as though the observer was looking at the celestial sphere "from the outside looking in", which is absolutely impossible. The figures appeared inside out, from a divine point of view at the time. This representation was initially used in celestial globes, which historically predate maps.

From the seventeenth century, the atlases of the sky began to appear with more precise and detailed mapping. The outside-in representation was gradually abandoned, and celestial maps became the right way round during the eighteenth century. The designs of the celestial figures, human or otherwise, varied considerably from author to author. The positions in which they were represented, the shapes, the complexions and the clothing (or lack of it) differed from map to map.

From the nineteenth century, decorative designs of the constellations were abandoned, in favour of greater precision in the marking of positions of the stars. The elaborately drawn figures had created confusion among astronomers, who above all wanted more detailed and informative maps, rather than beautiful designs. Modern day atlases include tens of thousands of stars, thousands of nebulae, galaxies and other celestial objects. All constellations are delimited by their borders, which were internationally agreed in 1930.

The Hindus and the Chinese developed their own constellations independently, according to their heroes and scenes from daily life. These are vastly different from the constellations derived from Greco-Roman mythology that are in use today: Chinese celestial maps had as many as 283 constellations. These representations are not internationally accepted, nor are their names. It is therefore not worth describing them here.

In this appendix, you can find, for the purposes of information only, some representations of ancient celestial figures with well-proportioned drawings, in relation to the imagined figures. Note that the relative positions of the stars are usually fairly rigorous. The figures have been adapted in order to make them easier to interpret and compare with current maps. The representations in Hevelius' (1611–87) atlas, published posthumously in 1690, were originally inside-out, but have been rearranged here so as to match what can be seen in the sky. Also included are adaptations of the figures in the most famous atlases of the seventeenth, eighteenth and nineteenth centuries. In all cases, the top figure is the ancient representation. I have put these together with current maps and borders of the constellations concerned, so that you can compare and draw your own conclusions as to the similarity or otherwise between the stars in the sky and the imagined figures. I have superimposed connecting lines between the brightest stars in the constellation, in order to make it easier to identify and compare the figures. Please remember that these lines are not, of course, part of the constellation. The modern representations, below the ancient figures, include all of the stars visible to the naked eye, in a good viewing location.

Ursa Major

Representation of Ursa Major, adapted from Hevelius' atlas, 1690. The seven most notable stars in this constellation, connected by broader lines in the bottom left hand drawing, make a shape which has different designations, according to local tradition: The Plough, The Big Dipper or The Saucepan. Ursa Major is in this position at the beginning of summer nights.

Figure A.2.

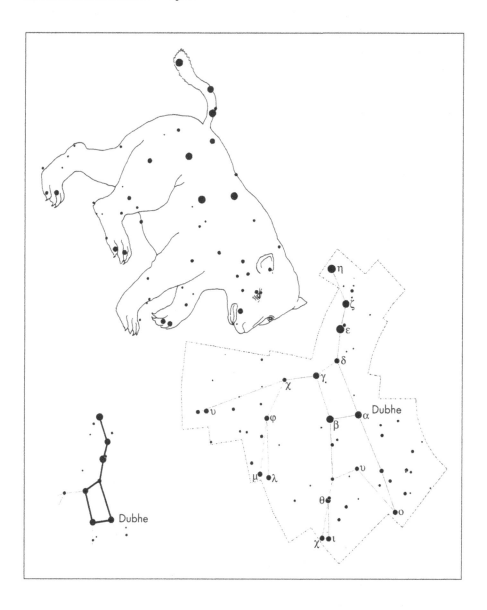

Cassiopeia

Representation of Cassiopeia, adapted from German lawyer and astronomer Johann Bayer's star atlas, 1603. This great atlas, known as *Uranometria*, contained 2000 stars. The five brightest stars in this constellation, connected by broader lines in the bottom right hand drawing, make a W (or an M) in the sky, which is not difficult to locate. Cassiopeia appears in this position at the beginning of summer nights.

Figure A.3.

Orion

The constellation Orion, the Hunter, adapted from Johann Bode's star atlas, *Uranographia*, 1801. This atlas contained over 17,000 stars and was the last great atlas to combine artistic figures with scientific representation. Three of the stars in this constellation, connected by broader lines in the bottom left hand drawing, are commonly known as The Three Marias or The Three Wise Men and depict Orion's Belt. Orion appears in this position on winter nights. He is defending himself against the Bull, using a lion's skin as a shield.

Figure A.4.

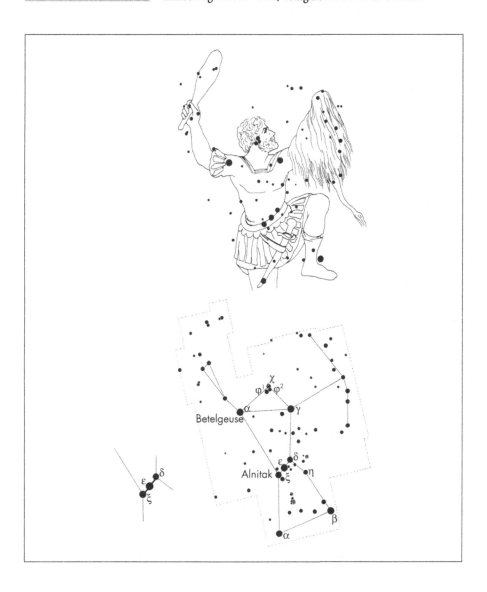

Taurus

Representation of Taurus, adapted from Johann Bayer's star atlas, 1603. Note that the star Aldebaran, bright and red, depicted one of the bull's eyes. This animal was represented attacking Orion, the Hunter.

Figure A.5.

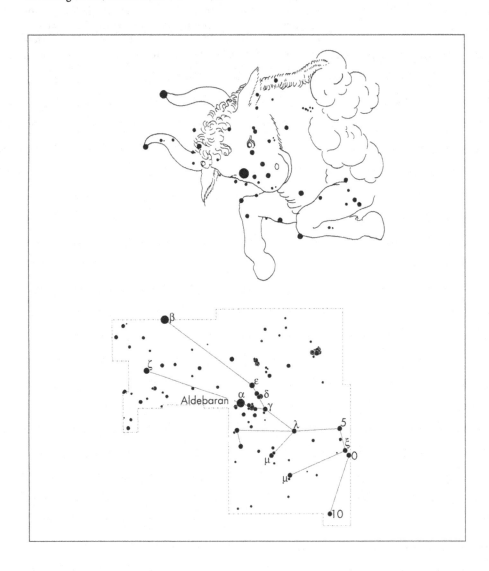

Centaurus, Lupus and the Southern Cross

Representation of Centaurus, Lupus and The Southern Cross, adapted from Hevelius' atlas, 1690. Centaurus, mythical half-man, half-horse, was imagined fighting with Lupus, the wolf, sticking his lance violently into the beast's neck. The Southern Cross can be seen at the feet of the Centaurus. The Southern Cross is the smallest of all the constellations, but it is also the most famous in the celestial southern hemisphere.

Figure A.6.

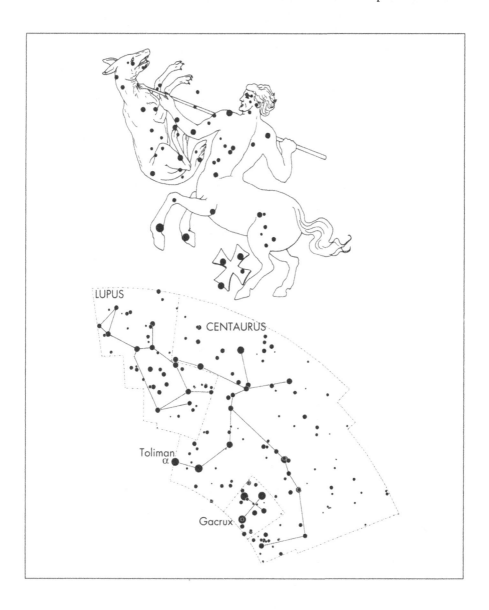

Andromeda, Pegasus and Triangulum

Representation of the constellations Andromeda, Pegasus and Triangulum adapted from Johann Bode's star atlas. The figure of the princess Andromeda, daughter of the King and Queen of Ethiopia (Cepheus and Cassiopeia) appears chained to a rock, or so according to the legend. The small constellation of Triangulum lies to the south of Andromeda. The star Sirrah (Andromeda), along with three conspicuous stars in Pegasus (Algenib, Markab and Scheat) make a prominent shape known as The Great Square of Pegasus. This is shown in broader lines in the bottom left hand drawing.

Figure A.7.

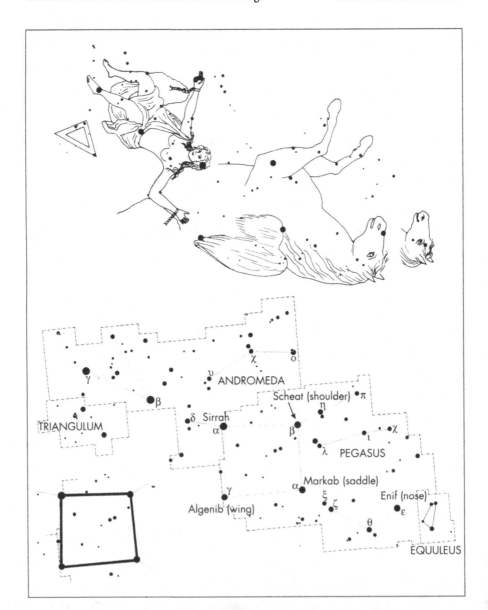

Scorpius

Figure A.8.

Representation of the constellation Scorpius adapted from Johann Bode's star atlas of 1801. The bright red star Antares represented the Scorpion's heart. Scorpius appears in this position at the beginning of July nights and only appears slightly above the southern horizon.

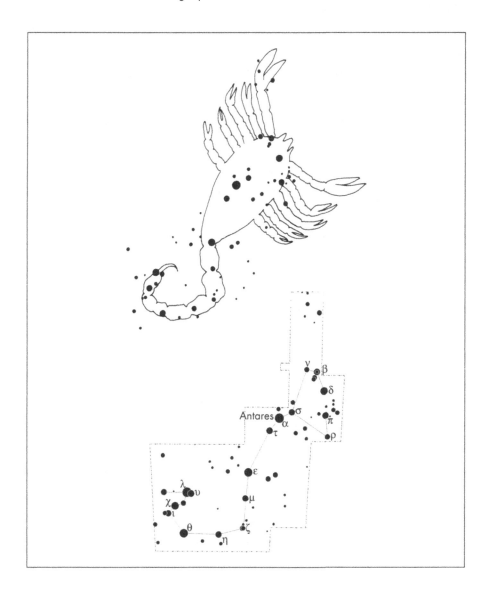

Sagittarius

Representation of Sagittarius, adapted from Hevelius' star atlas, 1690. The brightest stars in this constellation suggest a figure, which is nowadays referred to as a teapot (represented on the left). The similarity with the teapot is abundantly clear to our modern eyes. Next to Sagittarius' front legs is the constellation Corona Australis, the Southern Crown.

Figure A.9.

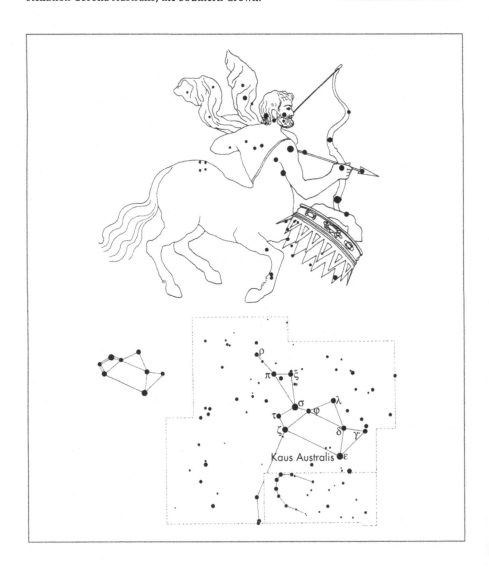

Leo

Representation of Leo, adapted from Hevelius' star atlas, 1690. Some stars of the lion's head and mane, connected by broader lines in the bottom right-hand drawing, form the shape of a scythe. Some people say that this figure is more like a back-to-front question mark (⸮).

Figure A.10.

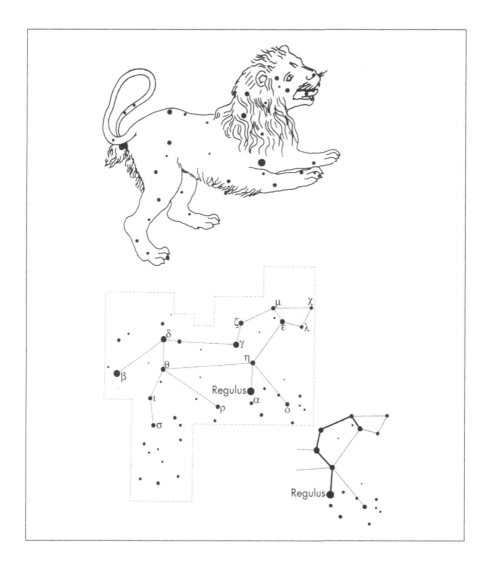

APPENDIX TWO

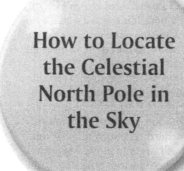

How to Locate the Celestial North Pole in the Sky

Figure A.11.

It has already been mentioned that the North Star (Polaris) is approximately in line with the celestial north pole, around 0.8° away from the pole. We have also seen how to locate Polaris (Figures 1.8 and 4.3). Let us now have a look at how to find the celestial north pole from Polaris (Fig. A.11). Imagine a line drawn in the sky from *Polaris* to Alkaid, the star that marks the tip of Ursa Major's tail. The celestial north pole is virtually on this line. Moving down this line an angular

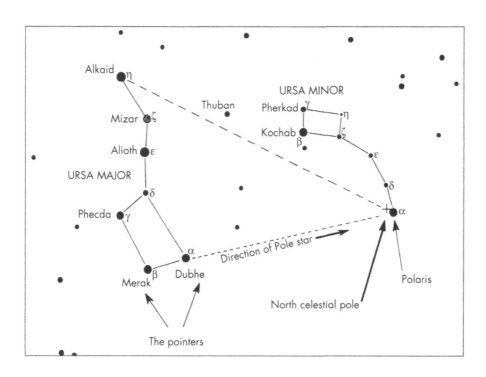

distance of around the width of a pencil held at arm's length, 0.8°, you find the celestial north pole. It is marked with a + on the diagram. Depending on when the observation is taking place, Polaris may be above, below, to the left or to the right of the celestial pole.

The positions of the celestial poles are not fixed. They alter extremely slowly over the course of the millennia, due to the precession movement of our planet. Currently, the direction of the axis of the Earth's rotation points almost towards Polaris, moving ever closer to the direction of this star until the year 2100, whereupon it will begin to move away. After this, other stars will gradually become the "pole stars". The complete cycle takes around 26,000 years.

In approximately 2800 BC, the axis of the Earth's rotation pointed in a direction very close to the star Thuban (α Draconis), shown in Identification Map 1 (and also in Figure A.11). This star, located around halfway between Pherkad (γ Ursae Minoris) and Mizar (ζ Ursae Majoris), was at that time the North Star.

In the year 14,000 the celestial pole will be located at around 3° from the star Vega (α Lyrae), which will be the North Star in this distant future.

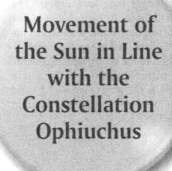

Movement of the Sun in Line with the Constellation Ophiuchus

The Sun passes in line with 13 constellations, as I said earlier when discussing the ecliptic. Figure A.12 shows the successive positions of the Sun, with a one-day interval, as the Sun passes in line with Libra, Scorpius, Ophiuchus and Sagittarius. As you can see, the section of the ecliptic in

Figure A.12.

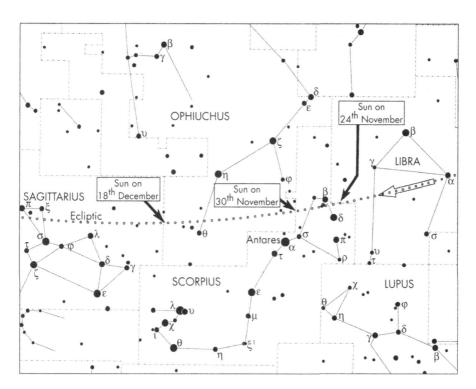

Ophiuchus is substantially longer than in Scorpius. The borders of the various constellations are also shown in fine lines. The dates indicated correspond to the positions of the Sun in relation to the constellations (and not to the signs).

The complete representation of the ecliptic, including the dates corresponding to the various positions of the Sun in relation to the constellations, is shown in Figures 3.27 and 3.28. The dates of the Sun's movement in relation to the signs of the zodiac can be found in Table 3.3.

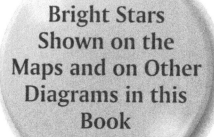

Bright Stars Shown on the Maps and on Other Diagrams in this Book

Using this appendix, the reader can find extra information on any star shown on a map or diagram in this book. The stars are listed in alphabetical order. The stars marked with an asterisk are not visible from the central belt of the USA or from Europe (Canopus is visible from Florida and Texas). Double, triple and multiple stars are shown as "d", "t" and "m", respectively, in the "characteristics" column. Their magnitude refers to that of the whole. The stars of variable brightness are shown as "v" in the "characteristics" column, the magnitude referring to the maximum brightness. The greater the distance the star is from us, the less precise we can be about that distance.

Star	Constellation	Apparent magnitude	Colour	Brightness (Sun = 1)	Distance (light years)	Characteristics
Achernar*	Eridanus	0.6	white	1000	144	d
Acrux*	Crux (Southern Cross)	0.9	white	3900	120	d
Adhara	Canis Major	1.6	blue-white	3500	430	giant, d
Albireo	Cygnus	3.1	yellow	690	390	giant, d
Alcor	Ursa Major	4.0	white	13	80	
Aldebaran	Taurus	0.9	reddish	145	65	giant
Alderamin	Cepheus	2.4	white	19	49	
Algenib	Pegasus	2.8	blue-white	620	330	
Algieba	Leo	2.3	gold-yellow	190	130	d
Algol	Perseus	2.2	blue-white	93	95	v (eclipse)
Alhena	Gemini	1.9	blue-white	140	105	
Alioth	Ursa Major	1.8	white	99	81	
Alkaid¹	Ursa Major	1.9	blue-white	140	100	
Almach	Andromeda	2.2	yellow	1400	355	t
Alnair	Grus	1.7	blue-white	160	100	
Alnilam	Orion	1.8	blue-white	25,000(?)	1200(?)	Supergiant
Alnitak	Orion	1.9	blue-white	19,000(?)	1100(?)	m
Alphard	Hydra	2.2	orange	400	180	Giant
Alrischa	Pisces	3.9	blue-white	44	140	d
Altair	Aquila	0.8	white	11	17	
Aludra	Canis Major	2.4	blue-white	52,500(?)	2500(?)	Supergiant
Ankaa	Fornax	2.4	yellow	50	78	Giant
Antares	Scorpius	1.0	red	10000	600	Supergiant, d
Arcturus	Boötes	0.0	orange	100	36	Giant
Arneb	Lepus	2.6	white	11,000(?)	1300(?)	Supergiant
Atria*	Triangulum Australe	1.9	reddish	2000(?)	400(?)	Giant
Avior	Carina	1.9	yellow	5500	630	Giant
Bellatrix	Orion	1.7	blue-white	1000	240	Giant
Betelgeuse	Orion	0.4	reddish	10,000	350	Supergiant, v

Star	Constellation	Apparent magnitude	Colour	Brightness (Sun = 1)	Distance (light years)	Characteristics
Canopus*2	Carina	-0.7	yellowish-white	13 000	310	Supergiant
Capella	Auriga	0.1	yellow	126	42	d
Caph	Cassiopeia	2.3	white	28	55	
Castor	Gemini	1.6	blue-white	47	52	m
Cih	Cassiopeia	2.5	blue-white	4 000	600	v
Cor Caroli	Canes Venatici	2.9	blue-white	65	110	d
Dabih	Capricornus	3.1	yellow	500(?)	340(?)	d
Deneb	Cygnus	1.3	blue-white	250 000	3200	Supergiant
Deneb Algiedi	Capricornus	2.9	white	8	39	d
Denebola	Leo	2.1	white	14	36	Na cauda do leSo
Diphda	Cetus	2.0	yellow	110	96	d
Dubhe	Ursa Major	1.8	yellow	220	124	Giant, d
Elnath	Taurus	1.7	bluish	290	130	Giant
Eltanin3	Draco	2.2	orange	210	150	Giant
Enif	Pegasus	2.4	yellow	3 900	670	Supergiant, triple
Errai	Cefeu	3.2	yellow	8	45	
Fomalhaut	Piscis Austrinus	1.2	blue-white	16	25	
Gacrux*	Crux (Southern Cross)	1.6	red	136	88	Giant
Gemma4	Corona Borealis	2.2	blue-white	55	75	d
Gienah	Corvus	2.6	blue-white	190	165	
Hadar*5	Centaurus	0.6	blue-white	12 000	525	
Hamal	Aries	2.0	yellow	52	66	Giant
Izar	Boötes	2.5	orange	380	210	Giant
Kaus Australis	Sagittarius	2.0	blue-white	310	145	Giant
Kochab	Ursa Minor	2.1	orange	180	126	
Markab	Pegasus	2.5	blue-white	150	140	Giant
Markeb	Vela	2.5	blue-white	2 300	540	
Megrez	Ursa Major	3.3	white	24	81	
Menkalinan	Auriga	2.1	white	90	82	d

Star	Constellation	Apparent magnitude	Colour	Brightness (Sun = 1)	Distance (light years)	Characteristics
Menkar	Cetus	2.5	reddish	360	220	Giant
Merak	Ursa Major	2.4	white	56	80	
Menkent	Centaurus	2.1	yellowish-white	42	61	
Miaplacidus*	Carina	1.8	blue-white	200	110	
Mimosa*	Crux (Southern Cross)	1.3	white	3000	350	
Mintaka	Orion	2.2	blue-white	10,000	1200	
Mira	Cetus	3.0	reddish	500(?)	420(?)	v (supergiant)
Mirach	Andromeda	2.1	reddish	450	200	Giant
Mirfak	Perseus	1.9	yellow	5100	590	Supergiant
Mirzam	Canis Major	2.0	blue	7000	500	Giant
Mizar	Ursa Major	2.3	white	60	78	m
Naos	Puppis	2.3	blue-white	20,900	1400	
Nunki	Sagittarius	2.1	white	600	220	
Peacock*	Pavo	1.9	blue-white	430	190	
Phecda	Ursa Major	2.5	white	58	84	
Pherkad	Ursa Minor	3.1	blue-white	1100	480	
Pole Star	Ursa Minor	2.1	yellow	2300	430	Supergiant, d
Pollux	Gemini	1.1	orange-yellow	30	34	Giant
Porrima	Virgo	2.8	blue-white	9	37	d
Procyon	Canis Minor	0.4	yellowish-white	7	11.4	d
Rasalgethi	Hercules	3.0	reddish	860	380	d, v (irregular)
Rasalhague	Ophiuchus	2.1	white	25	47	
Rastaban°	Draco	2.8	yellow	750	360	Supergiant
Regor	Vela	1.8	blue-white	11,000	900	m
Regulus	Leo	1.4	blue-white	131	77	d
Rigel	Orion	0.1	blue-white	38,000	770	Supergiant, d
Ruchbah	Cassiopeia	2.7	blue-white	65	100	
Sabik	Ophiuchus	2.4	blue-white	58	83	
Sadalmelek	Aquarius	3.0	yellow	3000	800	Supergiant

Star	Constellation	Apparent magnitude	Colour	Brightness (Sun = 1)	Distance (light years)	Characteristics
Sadalsud	Aquarius	2.9	yellow	2000	700	Supergiant
Sadr	Cygnus	2.2	yellow	22,000	1,400	Supergiant
Saiph	Orion	2.1	blue-white	7500	750	Supergiant
Scheat	Pegasus	2.4	reddish	320	200	Giant, v
Shaula	Scorpius	1.7	blue-white	8000	600	
Shedar[7]	Cassiopeia	2.2	yellow	500	240	Giant, d
Sheratan	Aries	2.6	white	24	60	
Sirius	Canis Major	-1.5	white	21	8.6	d
Sirrah[8]	Andromeda	2.2	blue-white	110	97	
Spica	Virgo	1.0	blue-white	2100	260	d
Suhail	Vela	2.2	yellow	3200	550	Supergiant
Thuban	Draco	3.7	white	250	310	
Toliman[9]	Centaurus	-0.3	yellow	1.5	4.3	t
Tureis[10]	Carina	2.3	yellowish-white	4800	650	Supergiant
Unukalhai	Serpens	2.7	orange	3673		Giant
Vega	Lyra	0.0	white	47	25	
Vindemiatrix	Virgo	2.8	yellow	58	100	Giant
Wezen	Canis Major	1.9	yellowish-white	45,000	2,000	Supergiant
Zosma	Leo	2.6	blue-white	24	58	
Zubeneschamali	Libra	2.6	greenish-yellow	18	160	

1 Also known as Benetnasch.
2 In Greek mythology, Canopus was the name of the pilot of the Menelau fleet, the Spartan king. Curiously, this star is now used as a guide in space navigation.
3 Or Etamin
4 Or Alphecca
5 Also known as Agena
6 Or Alwaid
7 Also known as Schedar, or Shedir
8 Or Alpheratz
9 Also known as Rigil Kentauris and Alpha Centauri
10 Or Aspidiske
11 Uncertain information (relatively high uncertainty).

Important note:

All of the stars mentioned in this appendix are brighter than our Sun. This should not, however, lead you to believe that our Sun is one of the least luminous. One should also not hastily deduce that most stars are giant or supergiant. In fact, the stars are very far away and so in order to be visible to the naked eye, they have to be extremely luminous in themselves. This is the only way they can be seen from afar; great distances "select" stars. Many others, of low or medium brightness, are not visible to the naked eye.

If the Sun were 57 light years away from us it would be at the limit of visibility to the unaided eye. At 400 light years, you would not even be able to see it with binoculars. Of the 14 stars that are closest to us, and which are thus closest to the Sun, only two are actually brighter than the Sun – Toliman (α Centauri) and Sirius (α Canis Majoris). The remaining 12 are considerably fainter. Of these 14 stars only five are clearly visible to the naked eye.

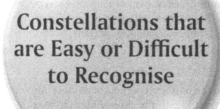

Constellations that are Easy or Difficult to Recognise

As mentioned in the first few pages, the constellations are not all equally easy to recognise. This degree of difficulty is to some extent subjective since, in addition to the characteristics of each constellation, it depends on the viewing location and the observer's eyesight and experience. Even so, here are some indications as to the level of difficulty of recognising constellations.

The division into four levels of difficulty is of course purely arbitrary. In each of these levels, some constellations are easier to recognise than others. The names are listed in alphabetical order. Appendix 6 offers further useful indications regarding the 88 constellations currently recognised and shows which are visible from places at latitudes close to 40° N.

Constellations that are conspicuous and stand out in the sky:
Aquila, Boötes, Cassiopeia, Centaurus, Cygnus, Auriga, Southern Cross, Scorpius, Gemini, Leo, Orion, Perseus, Carina, Sagittarius, Taurus, Ursa Major.

Constellations that are easy to recognise:
Andromeda, Canis Major, Canis Minor, Aries, Grus, Lyra, Lupus, Ophiuchus, Pegasus, Puppis, Triangulum Australe, Ursa Minor, Virgo, Vela.

Constellations that are relatively accessible:
Ara, Aquarius, Apus, Libra, Cetus, Circinus, Canes Venatici, Capricorn, Cancer, Corona Borealis, Corvus, Delphinus, Dorado, Draco, Eridanus, Phoenix, Hercules, Hydra, Hydrus, Indus, Lepus, Musca, Octans, Pavo, Piscis Austrinus, Pisces, Volans, Columba, Serpens (tail), Serpens (head), Reticulum, Sagitta, Tucana, Triangulum, Monoceros.

Very faint constellations, difficult to recognise:
Coma Berenices, Chamaeleon, Caelum, Pyxis, Corona
Australis, Scutum, Sculptor, Fornax, Camelopardalis, Lacerta,
Leo Minor, Lynx, Antlia, Mensa, Microscopium, Pictor,
Equuleus, Vulpecula, Norma, Horologium, Sextans, Crater,
Telescopium.

Note:
Some constellations in the bottom level do not possess any
stars above fourth magnitude. In the case of Mensa, none
reaches above fifth. However, the main stars in these constellations are shown on the Identification Maps, along with their
corresponding place in the celestial sphere. In the case of
Mensa, only its position in relation to neighbouring constellations is shown.

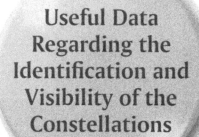

Useful Data Regarding the Identification and Visibility of the Constellations

This appendix shows the number of the Identification Map where each constellation can be found, as well as its visibility and location in the celestial sphere: north – celestial northern hemisphere; south – celestial southern hemisphere and equator – constellation which the celestial equator crosses. The last column shows the number of stars in the constellation with a luminosity above fourth magnitude. The names of the constellations are listed alphabetically. Appendix 5 gives the degree of difficulty in identifying the constellations.

Constellation	Identification map	Visible from latitude 40°N	Location	No. of stars over fourth mag
Andromeda	4, 5	Yes	North	14
Antlia (Air Pump)	2, 3	Yes	South	1
Apus	8	No	South	3
Aquarius (Water Bearer)	5, 6	Yes	Equator	16
Aquila (Eagle)	6, 7	Yes	North	11
Ara (Altar)	6, 7, 8	No	South	8
Aries (Ram)	4, 5	Yes	North	4
Auriga (Charioteer)	3, 4	Yes	North	9
Boötes (Herdsman)	1, 2, 7	Yes	North	15
Caelum (Chisel)	3, 4	Yes	South	2
Camelopardalis (Giraffe)	8	No	South	3
Camelopardalis (Giraffe)	1, 4	Yes	North	3
Cancer (Crab)	3	Yes	North	4
Canis Major (Great Dog)	3	Yes	South	19
Canis Minor (Little Dog)	3	Yes	North	2
Canes Venatici (Hunting Dogs)	2	Yes	North	2
Capricornus (Sea Goat)	5, 6	Yes	South	9
Carina (Keel)	3, 8	No	South	12
Cassiopeia	1, 4	Yes	North	10
Centaurus (Centaurus)	2, 7, 8	Partial	South	25
Cepheus	1, 5, 6	Yes	North	11
Cetus (Sea Monster)	4, 5	Yes	Equator	14
Circinus (Compasses)	8	No	South	2
Columba (Dove)	3	Yes	South	6
Coma Berenices (Berenice's Hair)	2	Yes	North	2
Corona Australis (Southern Crown)	6, 7	Yes	South	2
Corona Borealis (Nothern Crown)	7	Yes	North	6
Corvus (Crow)	2	Yes	South	6
Crater (Cup)	2	Yes	South	3
Crux (Southern Cross)	8	No	South	6

Constellation	Identification map	Visible from latitude 40°N	Location	No. of stars over fourth mag
Cygnus (Swan)	1, 5, 6, 7	Yes	North	23
Delphinus (Dolphin)	5, 6, 7	Yes	North	4
Dorado (Swordfish)	3, 8	No	South	14
Draco (Dragon)	1, 6, 7	Yes	North	15
Equuleus (Foal)	5, 6	Yes	North	2
Eridanus (River)	3, 4, 5, 8	Partial	South	28
Fornax (Furnace)	4	Yes	South	2
Gemini (Twins)	3	Yes	North	19
Grus (Crane)	5, 6, 8	Partial	South	9
Hercules	6, 7	Yes	North	20
Horologium (Clock)	4	Partial	South	1
Hydra (Sea Snake)	2, 3	Yes	Equator	16
Hydrus (Sea Serpent)	3	No	South	5
Indus (Indian)	5, 6, 8	No	South	3
Lacerta (Lizard)	1, 5, 6	Yes	North	2
Leo (Lion)	2, 3	Yes	North	16
Leo Minor (Little Lion)	2	Yes	North	3
Lepus (Hare)	3, 4	Yes	South	11
Libra (Scales)	2, 7	Yes	South	6
Lupus (Wolf)	2, 7, 8	Partial	South	13
Lynx (Lynx)	1, 3	Yes	North	14
Lyra (Lyre)	6, 7	Yes	North	8
Mensa (Table)	8	No	South	0
Microscopium (Microscope)	5, 6, 8	Yes	South	0
Monoceros (Unicorn)	3	Yes	Equator	6
Musca (Fly)	8	No	South	5
Norma (Square)	7	Partial	South	0
Octans (Octant)	8	No	South	2
Ophiuchus (Serpent Bearer)	6, 7	Yes	Equator	17
Orion	3, 4	Yes	Equator	28

Constellation	Identification map	Visible from latitude 40°N	Location	No. of stars over fourth mag
Pavo (Peacock)	8	No	South	10
Pegasus (Winged Horse)	5, 6	Yes	North	15
Perseus	1, 4	Yes	North	23
Phoenix	4, 5, 8	No	South	10
Pictor (Easel)	3, 8	No	South	2
Pisces (Fishes)	4, 5	Yes	North	7
Piscis Australis	5, 6	Yes	South	5
Puppis (Stern)	3, 8	Partial	South	7
Pyxis (Compass)	3, 8	Yes	South	1
Reticulum (Reticle)	8	No	South	2
Sagitta (Arrow)	5, 6, 7	Yes	North	4
Sagittarius (Archer)	6, 7	Yes	South	20
Scorpius (Scorpion)	6, 7	Yes	South	19
Sculptor	4, 5	Yes	South	2
Scutum (Shield)	6, 7	Yes	South	2
Serpens (Serpent)	6, 7	Yes	Equator	13
Sextans (Sextant)	2	Yes	Equator	0
Taurus (Bull)	3, 4	Yes	North	21
Telescopium (Telescope)	6, 7	Partial	South	1
Triangulum Australe (Southern Triangle)	8	No	South	4
Triangulum (Triangle)	4, 5	Yes	North	3
Tucana (Toucan)	8	No	South	6
Ursa Major (Great Bear)	1, 2, 3	Yes	North	23
Ursa Minor (Little Bear)	1	Yes	North	5
Vela (Sails)	2, 3, 8	Partial	South	8
Virgo (Maiden)	2, 7	Yes	Equator	16
Volans (Flying Fish)	8	No	South	6
Vulpecula (Little Fox)	5, 6, 7	Yes	North	1

The constellations partially visible from 40°N graze the southern horizon and, of course, do not rise any higher. This makes them even more difficult to observe and the already faint stars are invisible to the naked eye, due to the greater absorption of light by the Earth's atmosphere (see Section 2.8). In regions further south than 40°N, these constellations appear higher above the horizon, which improves their visibility appreciably.

Approximate Latitudes of some Cities in the Southern Hemisphere, or Close to the Equator

Country	City	Latitude
Australia	Adelaide	34° S
	Brisbane	27° S
	Cape Howe	37° S
	Darwin	12° S
	Hobart	43° S (Tasmania)
	Melbourne	37° S
	Perth	31° S
	Sydney	34° S
South Africa	Johannesburg	26° S
	Pretoria	25.5° S
New Zealand	Auckland	36°S
	Queenstown	45°S
	Wellington	41°S
United States of America	Houston	29.5° N
	Los Angeles	34.5° N
	Miami	25.5° N
	New Orleans	30° N

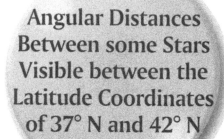

Angular Distances Between some Stars Visible between the Latitude Coordinates of 37° N and 42° N

The following table shows approximate angular distances between certain stars visible between 37° N and 42° N and other regions of the globe situated in the northern hemisphere. The angular distances shown are in ascending order. This table is useful for the reader to practise the hand method for measuring angular distances, as described in the beginning of the book and illustrated in Figure 3.21.

Between stars	Constellations	Angular distance
Atlas and Alcyone	Pleiades (Taurus)	0.38′ = 22.8°
North Star and celestial pole (north)	Ursa Minor/Ursa Minor	0.8°
Castor and Pollux	Gemini/Gemini	4.5°
Pointers of Ursa Major (Dubhe and Merak)	Ursa Major/Ursa Major	5.4°
Betelgeuse and Rigel	Orion/Orion	18°
Elnath and Betelgeuse	Taurus/Orion	22°
Deneb and Vega	Cygnus/Lyra	24°
Sirius and Procyon	Canis Major/Canis Minor	26°
Alkaid and Arcturus	Ursa Major/Boieiro	31°
Vega and Altair	Lyra/Aquila	34°
Deneb and Altair	Cygnus/Aquila	38°
Alkaid and Shica	Ursa Major/Virgo	61°
Antares and Regulus	Scorpius/Leo	100°

Example: the angular distance between Vega, the brightest star in Lyra, and Altair, the brightest star in Aquila, is 34.2º. This is a little more than a span + a fist, at arm's length (Fig. 3.21).

Another useful angular measurement: the apparent distance between two diametrically opposed points on the Moon, i.e. the Moon's apparent diameter, is approximately 0.5º.

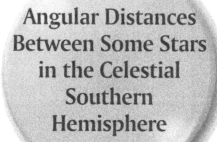

Angular Distances Between Some Stars in the Celestial Southern Hemisphere

This table shows approximate angular distances between some stars visible from places at latitudes considerably further south than 40° N, except for Sirius (α Scorpii) and Antares (α Canis Majoris), which are also visible from 40° N. Using this table the reader can practise hand measuring angular distances, as shown in Figure 3.21. Angular distances are shown in ascending order.

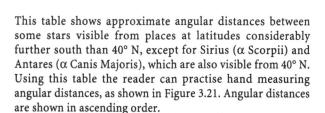

Between stars	Constellations	Angular distance
Mimosa and Gacrux	Southern Cross/Southern Cross	3.3°
Toliman and Agena	Centaurus/Centaurus	4.5°
Acrux and Gacrux	Stars of the Southern Cross	6°
Avior and Canopus	Carina/Carina	18°
Sirius and Canopus	Canis Major/Carina	36°
Antares and Toliman	Scorpius/Centaurus	39°
Acrux and Achernar	Southern Cross/Eridanus	59°
Alnair and Canopus	Grus/Carina	69°

Example: the angular distance between Sirius (α Canis Majoris) and Canopus (α Carinae) is 36°.

A further useful measurement: the angular distance between Acrux (α Crucis) and the celestial south pole is around 27°.

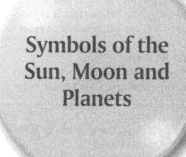

Symbols of the Sun, Moon and Planets

Alongside the symbols of the signs of the zodiac, as mentioned in Section 3.5 (Table 3.3), there are also symbols for the Sun, the Moon and the planets. These symbols were and are still used nowadays to represent the corresponding celestial bodies.

The planets Uranus, Neptune and Pluto were discovered relatively recently, so their symbols are more modern. The other symbols are ancient in origin. It is commonly believed that they date back to Greek or Roman mythology.

Below are the symbols used, followed overleaf by accompanying notes. Some of the symbols have variants, with small differences of detail.

Sun	Moon	Mercury	Venus
☉	☾	☿	♀

Earth	Mars	Jupiter	Saturn
♁ or ⊕	♂	♃	♄

Uranus	Neptune	Pluto
♅ or ⚇	♆ or ♆	♇

The symbol of the Sun probably depicts a shield with a protrusion in the centre. Some people believe that it represented the symbol of a palm-tree, used thousands of years ago in Egypt.

The Moon's symbol speaks for itself.

The symbol for Mercury represented caduceus, a winged laurel branch with two snakes entwined. This branch was attributed to the mythical god Mercury and was the insignia of ancient parliamentarians and heralds.

A hand-held mirror used by Venus, the goddess of love and beauty, is the symbol of the planet Venus.

The symbol for Mars is a lance pointing out from a shield. These were weapons of ancient warriors and were attributed to Mars, the mythical god of war.

The symbol for Jupiter is a hand-written Z, the first letter of Zeus, the Greek name for Jupiter, the father of the gods. There are those who believe that this symbol represented an imperfect hieroglyph for Aquila, the Eagle, the bird of Jupiter.

The symbol of Saturn is a misshaped representation of the scythe of time, insignia of Saturn, god of destiny.

Uranus was only discovered in 1781, by Sir William Herschel. This symbol is therefore very recent and represents the letter H, the first letter of Herschel, with the planet suspended from the horizontal bar of the letter.

Neptune was only discovered in 1846, due to calculations by Adams and Leverrier. Its symbol represents the trident of Neptune, god of the seas and oceans. Although recent, the symbol follows tradition.

Pluto was only discovered in 1930, by Clyde Tombaugh, although its existence was predicted by the calculations of Percival Lowell. As a tribute, the symbol of Pluto is a monogram based on the letters P and L.

Star Names

"Bayer Notation" is universally used to identify the brighter stars. The names are derived from the *Uranometria Star Catalog*, which was published in 1603.

Letters of the Greek alphabet are used in conjunction with the constellation name. Bayer listed the stars by brightness, with α (alpha) as the brightest star in a constellation, β (beta) the next, and so on to ω (omega). The Greek alphabet is given in Appendix 12, overleaf.

When writing Bayer names it is usual to use a lower-case Greek letter followed by the Latin version of the constellation name. The brightest star in the constellation of Ursa Minor (the Little Bear) is therefore "α Ursae Minoris". Note that "Ursa Minor" changes to "Ursae Minoris", to take the Latin genitive form, so that the name translates literally into "alpha of Ursa Minor".

Sometimes a three-letter abbreviation is used for the Greek letter, so you will often see "ALP Ursae Minoris". Equally often the constellation name is also abbreviated to three letters too, to give "α UMi" or "ALP UMi" (all meaning alpha Ursae Minoris).

Using the Bayer system makes it relatively easy to guess the identity of the brighter stars in every constellation, although unfortunately Bayer wasn't always consistent. For example Rigel – α Orionis – at magnitude 0.4, isn't as bright as Betelgeuse – β Orionis – at magnitude 0.15.

Which brings us to common and Arabic star names. In the USA and UK a few of the brightest or most unusual stars are also commonly known by their Arabic names; just a few of them because English-speakers generally find Arabic names confusing and hard to remember! Examples are Albireo (β Cygni) because it's a very prominent and beautiful double star, or Mira (o Ceti – that's omicron, by the way) because it is a regular naked-eye variable star.

A few common names are also in use, like the Pole Star (α UMi again), or the Dog Star (α CMa – it's the brightest star in Canis Major – see how it works?). But Bayer notation remains the easiest way to identify most stars.

The Greek Alphabet

The twenty-four letters of the classical Greek alphabet are given below.

Name	Lower case	Upper case
alpha	α	A
beta	β	B
gamma	γ	Γ
delta	δ	Δ
epsilon	ε	E
zeta	ζ	Z
eta	η	H
theta	θ	Θ
iota	ι	I
kappa	κ	K
lambda	λ	Λ
mu	μ	M
nu	ν	N
xi	ξ	Ξ
omicron	ο	O
pi	π	Π
rho	ρ	P
sigma	σ	Σ
tau	τ	T
upsilon	υ	Y
phi	φ	Φ
chi	χ	X
psi	ψ	Ψ
omega	ω	Ω